JOURNEYS FROM THE
CENTRE OF THE EARTH

JOURNEYS FROM THE CENTRE OF THE EARTH

IAIN STEWART

CENTURY

Published by Century in 2005

1 3 5 7 9 10 8 6 4 2

First published in the United Kingdom in 2005 by Century
The Random House Group Limited
20 Vauxhall Bridge Road, London SW1V 2SA

Random House Australia (Pty) Limited
20 Alfred Street, Milsons Point, Sydney,
New South Wales 2061, Australia

Random House New Zealand Limited
18 Poland Road, Glenfield
Auckland 10, New Zealand

Random House South Africa (Pty) Limited
Endulini, 5a Jubilee Road, Parktown 2193, South Africa

The Random House Group Limited Reg. No. 954009

www.randomhouse.co.uk

A CIP catalogue record for this book is available
from the British Library

Papers used by Random House are natural, recyclable products from wood grown in sustainable forests. The manufacturing processes conform to the environmental regulations of the country of origin

ISBN 1 8441 3813 5

Typeset in Bembo and Trajan
Printed and bound in Germany by Appl, Wemding
Artworks by Iain Stewart
Book design Dave Crook

To Paul Hancock, in whose footsteps I'm still treading

ACKNOWLEDGEMENTS

This book emerges out of a long-standing interest in how geology has impacted on the lives of people, both today and in the past. That interest was especially fostered in my early academic years by two mentors, Paul Hancock and Claudio Vita-Finzi, both of whom pursued research that strode roughshod across traditional disciplinary boundaries. As is so often the case, my geological world expanded greatly through the supervision of research students like Victoria Buck and Thomas Dewez (who also provided the illustration of the Mediterranean tectonic setting) and through collaboration with Suzanne Leroy, out of which developed our UNESCO/International Geological Correlation Program research initiative on Environmental Catastrophes and Human History. Much of the wide-ranging preoccupations of these individuals are buried in its pages, as is the work of numerous colleagues and fellow researchers who are hammering away at that exciting but tricky frontier of modern earth science that brushes up against history, archaeology, and anthropology. A casualty of writing a geology book for those that don't think they are interested in geology has meant the expunction of references to specific scientists, articles and books that fed this work, but I trust that those that feature here recognise their input and appreciate the dilemma.

The book started life first as the BBC/Discovery television series 'Journeys From The Centre Of The Earth' (or 'Hot Rocks' for US viewers). For that series I am indebted to Belinda Cherrington, who as Series Producer made it her baby, before leaving to produce a real one of her own (hi Rebekka). I am equally indebted to Jeremy Phillips who took over the challenge of making a popular mainstream geology series and saw it to its completion, aided by the enthusiastic and critical guidance of Clare Patterson and John Lynch. As well as thanks to those at BBC Specialist Factual that got the project off the ground - Cameron Bilbirnie, Michael Mosley, Steve Wilkinson and Matthew Barrett - I am grateful to all that made the filming of it such a fun time: Paul Olding, Nigel Miller, Nikki Meluish, Billie Pink, Kat Blair, Cathyrn O'Neill, George Hitchins, Nick Walker, Alan Parry, Jane Johnston, Patricia Fearnley, and especially my cameraman, Jeremy Humphries. Without their constant encouragement to persevere with the writing amidst our seemingly unremitting whistle-stop tour of Mediterranean airports and hotels this book would not have materialised. Neither would it have seen the light of day if it wasn't for the unflinching faith and support of my literary agent, Sheila Ableman, and the infectious enthusiasm of Christine King. At my publishers, Century, my idiosyncratic Scottish prose was gently whittled into shape by the editorial scalpel of Hannah Black and I'm grateful also to the design skills of Dave Crook.

Together, the series and book have taken more than a year and half out of my normal academic life. In this regard I owe a great debt to the help, patience and encouragement of many former colleagues in the Department of Geography and Earth Sciences at Brunel University, and in the Centre for Geosciences at the University of Glasgow, as well as current colleagues at the Department of Geology at Plymouth University. I can now get back to the day job!

But of course the greatest debts lie closer to home, with the friends and family who shared the excitement of the series and book unfolding, with the kids - Cara and Lauren - who for six months lost a Dad and gained an irritable itinerant lodger, and Paola - my wife - who kept everything on track, as ever. This book was their sacrifice and my glory.

CONTENTS

CHAPTER 1
MEDITERRANEAN WORLDS

In July 1831, something was disturbing the Mediterranean Sea off the south coast of Sicily. The waters churned and boiled, spouting great bursts of steam and ash high into the air, followed by thunderous eruptions of fire and cinders. There was the stink of sulphur, and all around dead fish floated on the surface of the sea. Far below, subterranean volcanic eruptions were pushing up a speck of sea bed until it broke above the waves – to become the Mediterranean's newest island, and a source of general astonishment.

It didn't take long for this tiny scrap of rock, less than two kilometres round and sixty-odd metres high, to cause an international political spat. The British saw its strategic value, close to major shipping lanes, and promptly planted a flag on it, naming it Graham's Island after Sir James Graham, then First Lord of the Admiralty. No sooner had the British departed than it was claimed by King Ferdinand II of the Two Sicilies (Italy was not yet united), and renamed after himself: Ferdinandea. The French and Spanish showed interest too, and the island became something of a tourist attraction. The diplomatic furore died only when everyone realised that, in the months they had been fighting over its sovereignty, it was sinking sheepishly back beneath the waves. By December it had disappeared from sight.

In recent years, Ferdinandea has been up to its old tricks. In 1987, the sub-sea volcanic mound had risen enough to be mistaken for a Libyan submarine by US aircraft, and unceremoniously depth-charged. By February 2000, fresh eruptions had brought the summit of the volcano to within eight metres of the sea surface, a threat to shipping. Fresh volcanic activity a year later led geologists to suggest that the next eruption would cause the island to re-emerge, opening again those old but unresolved diplomatic wounds. Meanwhile, the Italians have reportedly had a diver plant another flag on this subterranean territory, to remind rivals who owns it.

So: an unstable land, both geologically and politically – the history of the Mediterranean in miniature.

It's no fluke that Ferdinandea should occur where it did: no region more reflects the volatility of the Earth, or gives a better insight into its geology, than the Mediterranean. Not that this is the first thing about it that strikes most visitors. As one of the world's most popular holiday destinations, it attracts millions with its sun, sea and sand, its superlative food and wine, its rich art and culture. For many visitors, that's incentive enough. Then there are others who are in search of the past, treading ancient ruins to catch glimpses of past peoples and empires in this cradle of western civilisation, or the lure may be the wild landscapes that switch quickly from secluded island bays to magnificent high mountains, from seemingly lifeless desert wastelands to vast wetlands crammed full of birdlife.

Ironically, for all those holiday-makers drawn to the Mediterranean for relaxation, this is one of the most geologically violent places on Earth, with earthquakes, active volcanoes and even giant destructive tsunamis. Happening over human timescales, these natural phenomena have provided vital clues to understanding the way our planet works, an understanding that has founded the science of geology. Geology is about far more than just rocks: it's about understanding how the Earth works as a giant system – how the atmosphere, oceans, land surface and interior interact with each other and with the life on it, including ourselves. Just as we've learned to interpret the landscape, so its geology gives us insights into the human story of the Mediterranean.

INTERPRETING THE SIGNS

Throughout human history, people have only ever been able to interpret natural phenomena with the tools at their disposal, to the limits of an inquiring mind – and the constraints of prevailing dogma. So for most of our time on Earth we could have no true understanding of the forces we experienced. Our ancestors must have grappled with the mysteries of a capricious Nature, whose power was literally superhuman. Through observation, extrapolation and imagination, people constructed a world view that could give context to what they experienced – and in the process created mythological narratives that made us playthings of the gods. For, faced with something it cannot understand, what will the average human brain do? Impel its body to run away fast in the opposite direction, perhaps. Or, if danger doesn't seem immediate, it may just as likely start a process of thought, reflection – and imagination. Why does a rock look like a great big head, or a funny-shaped depression in the ground resemble a giant footprint? There must be a reason for the river drying up at one time of year, only to become a rushing torrent at another – surely a great animal must first be drinking thirstily, then spitting it all out . . . And when the sky darkens and the tumultuous air assumes ominous shapes that writhe in the gloom – those fierce bright jagged spikes that shoot out of the sky must be the bolts of a wrathful God . . . A mountain suddenly shakes and spews up masses of ash and steam way up into the sky and a fiery runny substance that flows down the slope burning all it touches – so the god who lives there must be angry. That rumbling noise we've heard for days must have been his hungry belly – we should have fed him but we didn't and now he's punishing us. We must propitiate him. Cue for sacrifice.

The evocative ruins of Delphi nestling in the majestic Parnassos mountains of Greece. In its heyday of Classical Greece, this ancient city was the religious centre of the known world. Here, dedications were made to a pantheon of gods and spirits, with the most lavish offerings being made to the renowned Delphic oracle, considered throughout the Mediterranean and Near East to be the supreme authority on how the world worked.

A rather simplistic view of the birth of a myth, no doubt, but there's no denying the huge range of human imagination when it comes to making a kind of sense of otherwise inexplicable phenomena – as the diverse civilisations of the ancient Mediterranean can amply testify. Today, to those who accept scientific explanations of the world, myths are just that: inventive, evocative stories that reflect a culture's understanding of its world with the means at its disposal, giving telling insights into human creativity, meeting a powerful need – a need that evidently persists today. Perhaps people have always responded to the fantastical, the other-worldly, something that transcends everyday reality. So even in our modern age there seems to be a widespread hunger for supernatural connections, anything from divine chariots to biblical codes to alien abductors. It seems that no amount of rational debunking will shake the faith of these devotees. Atlantis is a case in point: a legend from the remote past related by Plato 2,500 years ago, it is still a potent source of speculation. Plato's account of the mysterious island's cataclysmic destruction takes only a few hundred words, but this meagre foundation supports the huge weight of subsequent theories on which the modern mystery is based. Today, most of those theories are breathtakingly fantastic, and the 'lost world' has now been 'found' in places as disparate as the Caribbean, South America, Antarctica and Ireland.

To the early people of the Mediterranean, a whole pantheon of gods and spirits ordered their restless land, responsible for everything from drought and flood to famine and plenty. Naturally, these divine beings would have to be worshipped, placated, and sacrificed to. The Greeks had a particularly strong belief system, which makes it all the more remarkable that the first stirrings of scientific analysis arose in their culture. Early thinkers would search for more earthly explanations when catastrophes happened, and leave a record of those thoughts – none more so than the hugely influential philosopher Aristotle (384–322 BC), a pupil of Plato. He examined and defined the whole concept of logic, which he said was the only proper tool of any inquiry into the world around us. He created an ingenious model of rational thought: put simply, it centred on the idea that theory must follow fact. He applied his reasoning in many spheres of learning, from astronomy to politics to literary criticism, believing that each provided clues to the structure and order of the world.

Aristotle's contribution to geology was probably inspired by one of the natural disasters that his contemporaries would attribute to an angry god – the huge earthquake and tsunami that in 373 BC destroyed the Greek city of Helike, sending both city and inhabitants to the bottom of the sea. (It later became one of the contenders for being the 'real' Atlantis.) Helike led Aristotle to formulate a theory that earthquakes and accompanying seismic sea waves were the physical product of meteorological conditions rather than supernatural actions of gods. He came to the conclusion that earthquakes were caused by winds trapped in Earth struggling to get out, an idea that hung around for the next two thousand years.

Aristotle had also addressed the phenomenon of marine seashells within outcrops of rock, making it clear that part of what was now land had once been covered by sea. In his *Meteorologia*, he wrote of periodic switches between land and sea, but generally thought that they occurred too slowly or over too long a time interval for anyone to notice them happening.

The Greek philosophers like Aristotle popularised the notion of rational explanations of the natural world that was to influence thinkers for centuries to come. Observations of nature itself became the key to understanding how the world worked. Travellers like Strabo (63 BC–AD 25) and Pausanius (AD 120–180) wrote the first travel guides to the Mediterranean, describing in detail the physical world around them. Roman scholars like Seneca (AD

4–65) and Pliny the Elder (AD 23–79) produced weighty tomes on the natural world, while the latter's nephew Pliny the Younger (AD 61–113) gave the world the first detailed account of a volcanic eruption (that of Vesuvius, which took the life of his uncle). For most serious-minded scholars, what was controlling the cosmos was not a pantheon of higher beings but the competing elements of water, air, earth and fire. Meanwhile, most ordinary Greek and Roman folk still put their trust in a barrage of gods – whose powers, however, were about to be dramatically overturned.

■ A ROCKY ROAD TO KNOWLEDGE

When Constantine, a Christian convert, emerged as undisputed Roman Emperor in AD 312, Christianity became the dominant religion of western Europe; the Bible now provided a framework within which the world could be comprehensively explained. The planet had been created by God for mankind; there was no need to look for natural origins for catastrophes like earthquakes, volcanic eruptions and floods, or to debate which deity had been upset – all were down to the single Christian God. In fact, looking to nature to explain what was clearly God's holy design was a dangerous business. Number 102 in a list of over 150 heresies that one Roman bishop compiled in the late fourth century AD was belief in a natural origin for earthquakes, on the grounds that it was a negation of the power of God and a product of a pagan culture. For the next thousand years, no one would pay too much attention to geology.

However, the Renaissance of Europe in the fifteenth and sixteenth centuries heralded the birth of the modern, rational age. Just as ancient thinkers had begun to look beyond pagan dogma, so certain Christian scholars began to grapple with reconciling biblical teaching with the world they saw around them. One bone of contention became the age of the Earth. The Bible stated that God had created the planet in seven days, but how long were the 'days'? In the seventeenth century, Isaac Newton, who included the chronology of the ancient world among his many interests, suggested that at the beginning the Earth used to rotate more slowly, which would explain why it could be made in a week. A few decades later James Ussher, the Archbishop of Armagh in Ireland, used the Bible to calculate when the world began. He came up with 23 October 4004 BC – in fact, God started making it at noon on that day. (For the Jewish calendar introduced by Hillel II in AD 359, the starting date for the world had been 3761 BC.) Even Newton agreed with Ussher, and his date became widely accepted for the next two hundred years.

The towering basilica of St Peter on the west bank of Rome's river Tiber marks the heart of the Vatican City, home to the Holy Roman Catholic Church for over fifteen hundred years. When the Roman emperor Constantine adopted Christianity in the early fourth century AD, the Roman beliefs in a barrage of pagan gods were quickly overturned by a religion in which the world could be explained by the actions of a single Christian god.

One seed of doubt that nagged away at the minds of Christian scholars trying to fit the world into this new biblical timescale was that provided by fossils. In the 1500s, nearly two millennia after Aristotle's speculations, Leonardo da Vinci realised that fossils on the tops of hills were the shells of once living sea creatures and proof that the positions of the land and sea had changed. In the seventeenth century, the Danish scientist Nicolaus Steno, who specialised in the structure of the brain, wrote a geological 'anatomy' of the Tuscan landscape around him. His description of marine fossils on land, far from sea, demanded a much longer history than the 6,000 years suggested by the Bible. Steno was the first man to claim that Earth's past was chronicled in its layers of rock, thus displacing the Bible as the sole authority on the subject. He could have changed western civilisation's idea of history and nature and precipitated further investigation at this time; but he gave up geological studies, ironically to become a priest.

The apparent conundrum that the position of the lands and sea had changed was actually easily explained by the Bible. A fundamental belief of Christian dogma was that Earth had experienced only one major cataclysm in its natural history – the flood of Noah, dated by Ussher's chronology to have occurred in 2349 BC. In fact, the story of Noah is just one of many flood myths worldwide, most involving a man and a woman escaping by boat.

The Great Flood became the basis of very early geology: the mountains and landscape had all been carved out by one giant flood. Fossils, when accepted as the remains of dead organisms, simply became part of the debris of life left behind by the great deluge. However, people began to realise that forty days and nights of rain would not be sufficient to cover all the continents, and, although the biblical age of Earth was not openly refuted, most serious thinkers about the natural world quietly ignored Ussher's few thousand years, knowing the span to be significantly longer. Still pinning their ideas on some divine control, most took the view that God had set natural laws in motion at the start of creation, after which they operated like clockwork – but in the mid eighteenth century, the geology of the Mediterranean would spring a surprise that forced people to think again about how their world worked.

Fossils, such as this beautifully preserved sea shell embedded in limestone, were one of the crucial clues to understanding Earth's past. From antiquity onwards fossils had been unearthed all from the tops of mountains to the seashore, and by the 18th century they were accepted as the remains of once living creatures. It would be the differences in fossils in rock strata laid down at very different times in Earth history that would provide geologists with a tool for reconstructing the planet's history.

UNDERMINING THE FOUNDATIONS

By the mid eighteenth century, Lisbon was the capital of an empire rivalled only by Spain. Portuguese explorers had travelled the globe, bringing back the riches of the New World and the Orient. It had become one of the world's great cities. Then on 1 November, 1755, at 9.30 a.m., a catastrophe occurred that sent shock waves throughout Europe. In Lisbon there were three distinct earthquake shocks over ten minutes. The first was a warning signal. The second and third did most of the damage. Violent shaking was also experienced in north Morocco, Algiers and the Iberian peninsula. Even France, Italy and Switzerland felt the land shake. Lakes across Europe were disturbed.

In Lisbon there was devastation. It being All Saints' Day, the city's churches had been packed with worshippers, but the shaking brought the roofs down on Lisbon's great cathedrals. Few escaped alive and many of those who made it out to the streets were killed by falling masonry. Worse was to come. Candles that were lit in churches and homes throughout Lisbon overturned, and the ensuing fire ravaged the city for six days before it could be extinguished. People rushed to the harbour to escape the inferno but, thirty minutes after the earthquake, a sea wave six metres high that had been triggered by the quake far offshore now swept over the harbour and surged up the Tagus river, capsizing boats and drowning many who had fled to safety. All told, it is estimated that 70,000 people lost their lives in what is still the most disastrous earthquake to have struck Europe in recorded history.

The catastrophe had a profound impact. The very international nature of the city meant that first-hand accounts of the earthquake spread rapidly

throughout Europe, making it the first reported international disaster, and created a crisis of confidence in how widely accepted beliefs worked. Why would God destroy his own churches – and on such a holy day? Hard-line religious orders such as the Jesuits pronounced that the earthquake was God's wrath on the people of Lisbon for their sins. Others weren't so sure. In particular, the man who was essentially the dictator of Portugal, the Marques de Pombal, believed that the earthquake was a natural occurrence. His no-nonsense response to the aftermath of the quake was to bury the dead and feed the living, and he was furious with the Jesuits for stirring up religious fervour at this tragic time. In 1759, Pombal exiled the order, and over the next decade they were thrown out of France, Spain and many of the kingdoms of Italy.

Across Europe the rejection of the Jesuits reflected a seismic shift in thinking. Leading philosophers like Voltaire used the earthquake in an attempt to disprove divine providence in the world, claiming that God had abandoned Earth to men and could therefore no longer show his wrath to the world. Immanuel Kant spoke of a subterranean fire that had created the quake, while some even argued for a sudden contraction of the Earth.

Across southern Europe, the Lisbon earthquake forced a highly spiritual culture to decide between scientific, rational explanations and those presented by religious thinkers of the age. It wasn't exactly the end for God, but science was gaining momentum and a whole new way of thinking was emerging. However, for those who challenged the ideas of creation in Genesis, there was still the essential, burning question: 'How old is the Earth?' Disproving the Church's claim of 6,000 years was not going to be easy.

The roofless Convent de Carmelo in Lisbon, a casualty of the catastrophic earthquake that struck the city on 1st November 1755. The quake caused widespread collapse of churches and buildings, generated a devastating tsunami and started an enormous fire that reduced much of the city to ruins on one of the holiest days of the year. More than its physical impact, the earthquake created shockwaves in human thinking, as the idea gained ground that the earthquake was not an act of a wrathful God but a natural phenomenon.

A TOUR THROUGH TIME

The eighteenth century was a time of travel for people with sufficient money and leisure. The 'Grand Tour' became the young gentleman's rite of passage, an opportunity for aristocrats in northern Europe to broaden their horizons and learn about the politics, culture and art of neighbouring lands – rather like the modern gap year for students. For most it was a nostalgia trip through the lost world of the Classical era, visiting Athens, Venice, Florence and above all Rome, but it would be the young, restless geology of the Mediterranean that would capture the imagination of many of these first tourists, and it came to revolutionise the way we see our world.

George Louis Leclerc, known to posterity as the Comte de Buffon, was one of the century's nouveau riche. Forced to leave university in Dijon after becoming involved in a duel, Leclerc set off on the Grand Tour. He was so captivated by the Italian mountains that he went on to produce more than forty volumes full of observations on the natural world. He was particularly interested in finding out the age of the Earth. Leclerc had some unconventional theories, but avoided conflict with the Church with the time-honoured formula of presenting them as merely 'philosophical explanation'.

Leclerc believed that Earth originated as a molten body which, just like the other planets, had been broken off the sun by the impact of a comet. He proposed that its age could be calculated by estimating how long it took to cool down. He heated up different-sized balls of iron until they were white hot. He touched them at repeated intervals and timed how long they took to cool, no longer burning the skin. In 1778 he published his findings, estimating it would have taken at least 35,000 years to cool down sufficiently to allow condensation of atmospheric water vapour into a universal ocean. Further cooling over tens of thousands of years caused cracks to appear in Earth's surface to allow the seawater to drain down to its present level. As volcanoes erupted, the continents appeared and valleys were gouged out by ocean currents, before further cooling caused more erosion until the planet achieved its present form. In all, Earth was probably 75,000 years old. Leclerc's attempt at a scientific method of estimation challenged the Church's beliefs.

Just about all the geological knowledge then available to science was crammed into Leclerc's remarkable work, and it was phenomenally influential. Around that time the biblical ideas of the great deluge had led some geologists to develop a grand theory that most of the planet's rocks had originated under the waters of a 'universal' ocean. The view, popularised in the 1770s by the leading German geologist Abraham Werner, was called 'Neptunism' after the Roman god of the sea. Leclerc's 'Plutonist' view that internal fire was the process that shaped our planet was completely opposed to this, and support came from an unlikely source – a Scottish doctor, gentleman farmer and devout Christian. In the late 1780s, James Hutton argued that contorted rocks in the Scottish Highlands must have been the result of pressures exerted by great masses of molten rock. The retreat of the waters of the ocean was caused by the lands rising into mountains, not water being evaporated off into outer space as Werner believed. As the eighteenth century ended, a debate raged between these 'fire' and 'water' camps. Just as in the Classical era, scholars were again debating the fundamental physical elements of the cosmos.

Neither Werner nor Hutton ever went on the Grand Tour. Werner's theories were inspired by the rocks around his home in Freiberg, while Hutton relied for inspiration on his trips around Scotland and England and a quick visit to the Alps. Their theories about how the world worked were developing based on regions where geological actions were slow or extinct. Their followers who did embark on the Grand Tour were able to witness geology in action, and it soon became clear which of the two camps was correct. Those visiting in the Mediterranean were able to see young live volcanoes at Mount Etna and Mount Vesuvius, as well as their ancient extinct relatives in the Italian hills above Padua and Verona and in the Massif Central of southern France. Gradually the Neptunists conceded defeat, and the views of Leclerc and Hutton carried the day.

There was another battle to be won. The theories of both the Neptunists and Plutonists ran contrary to the official biblical line that Noah's flood was the principal force in shaping a planet with a very short history. Even Werner's waterworld was far older than a few thousand years, while Hutton saw the planet as being in a never-ending cycle – constantly renewing and destroying itself. Hutton recognised that rather than being shaped by cataclysmic forces, the planet was subject to small, gradual changes operating over enormous timescales. It was a direct challenge to those who held on to the Flood. Again two camps emerged, the 'diluvialists' who supported the flood, and the 'gradualists' who believed that all our planet needed was time. Inspired by the debate, a young Scots-born geologist headed off on the Grand Tour to put the case to rest.

Charles Lyell had started out dating strata by fossils to try to find geological evidence for the Flood, but in 1828 he arrived in Italy, and at Mount Etna in Sicily he found fossil deposits that had clearly built up gradually over long periods of time. Not only that, but they were elevated hundreds of metres above the sea where their modern equivalents still lived. It dawned on him that one of the largest piles of volcanic material in Europe had formed in a very recent geological time. Just how recent became clearer from a site near Naples. The Temple of Serapis in the harbour town of Pozzuoli on the Bay of Naples is actually an ancient Roman marketplace, but the stone columns here have dark bands of holes on them which Lyell realised were the borings of marine molluscs – and the molluscs were testament to a changing world. Because the Roman pillars were originally built above sea level, Lyell knew that since Roman times the columns must have sunk to become partially submerged in the sea, and then at some later time emerged with their molluscs to where they are today, several metres above sea level. He guessed correctly that the up-and-down movements were due to swelling and sinking of the hot volcanic rocks beneath the bay, but what was more amazing to Lyell was how much movement had occurred since Roman times. Hutton was right – gradual processes were going on, but the speed at which the Earth was capable of moving meant that with hundreds of thousands or even millions of years to play with, it could build mountains.

Just a couple of years later, the now-you-see-it-now-you-don't island of Ferdinandea was to be greeted as a major scientific prodigy. Charles Lyell would devote six pages to it in his influential book *The Principles of Geology* as further evidence that Earth was a dynamic place. What was sea could quickly become land, and vice versa. To add to the venerable accounts of lands rising from the sea and of seashells on mountainsides, here was proof that Earth's surface was in a constant state of renewal. The fleeting glimpse of Ferdinandea showed how quickly the planet could reshape itself.

Lyell deduced that geological processes operating gradually could, as Hutton had said, over time build mountains, valleys, canyons and all the other features we see today. To Lyell, the kinds of forces that affected our planet in the past must be assumed to have been exactly those we see in operation today – no faster, no slower – and no need for catastrophes. His theory, known as 'uniformitarianism', became the foundation stone of modern scientific thought. Lyell had done more than settle a debate, he had effectively turned geology into a science. Earth's past may be written in its rocks, but its processes are all around us today.

By the middle of the nineteenth century, our grasp of how the world worked had a firm scientific foothold, but we still didn't know how old it was. A hundred and fifty years after Steno realised that fossils could be used to build up a picture of the geological past, fossils became the tools to divide up Earth history. Three ages of life were defined: the age of new animal life (Cenozoic era), the age of middle life (Mesozoic era), and the age of ancient animal life (Palaeozoic era). Together these three eras made up the Phanerozoic aeon, the age of visible life. The Cenozoic era subsequently became divided into the Tertiary and the Quaternary periods. It would be Charles Lyell who, on the basis of the percentage of still-living species found in the rocks, would further divide the Tertiary into the Pliocene, Miocene and Eocene epochs.

Meanwhile other geologists had focused their attention on the older eras, splitting the Mesozoic into the Cretaceous, Jurassic and Triassic periods, while the Palaeozoic was broken into the Permian, Silurian and Cambrian periods. Tweaks and rearrangements of this scheme, often with bitter wrangling and controversy, would follow (and give two more Palaeozoic periods – Carboniferous and Devonian), but by 1850 this basic breakdown of the history of geological life on the planet was in place. Geologists had the pigeonholes in which they could place strata and fossils relative to each other – they just didn't know how old they were.

Columns in an ancient Roman marketplace at Pozzuoli near Naples in Italy provide clear signs of a restless Earth. A trio of huge limestone pillars in the so-called Temple of Serapis have dark bands around them where the rock has been riddled with holes from rock-chomping marine molluscs. In his visit to the site in the mid 19th century, Charles Lyell recognised that the columns must have subsided several metres to become submerged in the sea and attacked by molluscs, then raised several metres to their present elevation above sea level, all since Roman times. Geological processes were happening on human timescales.

Lyell had made a massive breakthrough, but he didn't just give birth to Geology; his work was the forerunner for Biology. When Charles Darwin embarked on his voyage in the *Beagle*, he took Lyell's ground-breaking book with him and used Lyell's encyclopedic knowledge of fossils as the basis for his theory of evolution. When asked about where he got the inspiration for his *The Origin of Species*, Darwin replied, 'I really think my books come half out of Lyell's brain. I see through his eyes.'

Ironically, while Hutton and Lyell offered no precise age for the Earth as a whole, preferring vague terms like 'unfathomable', Darwin at least had a stab at estimating its antiquity. He argued that erosion of southern Britain must have been going on for something like 300 million years, so the Earth itself must be far older than that. Twenty years later a British engineer, John Phillips, used this approach to estimate that the age of the planet itself was up to 600 million years. In the 1880s, the eminent physicist Lord Kelvin calculated the age on the basis of Earth cooling from a molten body, claiming it could not be more than 400 million years, but later, in 1899, he calculated his final estimate: a mere 24 million years.

Kelvin's calculation was way out. He had missed one crucial factor, mainly because it had only been discovered a few years earlier by the French physicist Henri Becquerel. Radioactivity – the heat that the planet itself generates through the steady breakdown of unstable elements – would counteract any heat loss from the cooling Earth. In 1904, Ernest Rutherford would highlight the role of radioactivity in maintaining the Earth's heat and estimate the planet's age to be hundreds if not thousands of millions of years. Today, it is generally accepted among Earth scientists that the planet is about four and a half billion years old and that organisms have been living on it for more than three billion years. Back at the dawn of the twentieth century these mind-boggling ages had yet to be worked out, but geologists finally had a timescale that fitted with the observations of the likes of Hutton and Lyell.

■ DRIFT TO A MODERN WORLD

It took another sixty years until we finally came to an understanding of what was really behind how the Earth beneath our feet actually works. During the First World War, in a university in Marburg, Germany, a 31-year-old meteorologist, Alfred Wegener, was browsing though the library. He came across a scientific paper that listed fossils of identical plants and animals found on opposite sides of the Atlantic Ocean. Intrigued by this information, Wegener began to find more cases of similar organisms separated by great oceans.

At the time it was believed that landbridges had once connected far-flung continents, but Wegener noticed the close fit between the continents of Africa and South America, and thought that the similarities between organisms could be explained by the fact that at one time all the continents had been joined together. Wegener soon began to find geological evidence to support his theory. He noticed that large-scale features on separate continents often matched very closely when the continents were put together. The Appalachian mountains of eastern North America matched with the Scottish Highlands, and the distinctive rock strata of the Karroo system of South Africa were identical to those of the Santa Catarina system in Brazil.

Wegener also began to notice that fossils found in certain places often indicated a climate utterly different from that of today. Using all the information he had collated, Wegener wrote *The Origins of Continents and Oceans*, proposing the theory of 'continental drift', centred on landmasses moving across the Earth's surface. As with many radical theories, Wegener's ideas at first provoked much hostility and were described as 'footloose' and reliant on 'ugly facts'. One American scientist was prompted to say, 'If we are to believe this hypothesis, we must forget everything we have learned in the last seventy years and start again.' It was true that his driving force for drift, Earth's rotation, was ridiculously insufficient and that his notion of continents floating like huge rafts on their ocean basement was too simple, but Wegener had proposed a moving world that explained why the landscape changed slowly. Within a few decades, his ideas turned from ridicule to orthodoxy when in the 1960s the ideas of drift were combined with new driving forces and - relabelled as plate tectonics (the focus of the following chapter). Plate tectonics was revolutionary in that it explained the world as a system of slow-moving parts, but in many respects plate tectonic theory was perfect uniformitarianism. A more serious threat to world order was just round the corner.

A NEW DOGMA

Geology had come a long way. By the 1970s we knew about the age of the Earth, the slowness of the processes that created it, and now the existence of mobile tectonic plates. It felt as if we had cracked it. Science, however, doesn't work like that. Just as one set of ideas and beliefs establishes itself, something big happens which forces us to think again, and the last big one was a quarter of a century ago.

While the Christian framework had left no room for processes taking a long time, the new framework of uniformitarianism left little room for the influence of catastrophic events. One dogma had been replaced by another. In 1977 Walter Alvarez, a Californian, was studying a thin band of reddish clay between two layers of limestone in the Botticcione Gorge in Umbria, Italy. The clay band was very important since it marks the division between two periods of time, the Tertiary and the Cretaceous, about sixty-five million years ago. This KT boundary (the 'K' in KT comes from the German for Cretaceous to avoid confusion with the Carboniferous) is the time when the dinosaurs and half the world's species died out. Alvarez wondered what clues this 6-millimetre-thin layer of clay offered for such a dramatic moment in the world's history. Luckily for him, his father had a Nobel Prize in physics and a few good connections, so the red clay was sent to a lab for analysis. When the results came back, everyone was astounded. It contained over 300 times the normal level of the element iridium.

Iridium is an exotic element, rare on Earth, where it is found only deep in the core, but abundant in space. The Alvarez group used the high iridium concentrations to suggest that 65 million years ago, a time when the limestones at Botticcione were the soft lime-rich muddy sea bed, a lump of rock about a kilometre in diameter had struck Earth. The disintegration of the rocky projectile had dusted the globe with a minute coating of the iridium-rich cosmic dust. Over the next few years, rock samples from the KT boundary sites all over the world were analysed and found to show similar greatly elevated levels of iridium. By 1984, most geologists were ready to accept that the Alvarezes had demonstrated a global catastrophe and that the massive extinction at the end of the Cretaceous period was caused by some extraterrestrial object.

The idea that the Earth could be subject to devastating impacts from outer space was not new: in the seventeenth century Edmond Halley, who identified the comet that bears his name, suggested that a projectile splashing down in the Caspian Sea might account for the story of Noah's Flood. This scientific proof that our planet, and the life on it, could be reshaped in an instant came as a shock to geologists brought up in a creed of uniformitarianism.

Today, debate over whether or not an impact from outer space alone was enough to cause this great KT extinction of life is still dragging on, but catastrophic processes have emerged out of the woodwork of our own planet, as geologists recognise the monstrous effects of supervolcanoes, giant earthquakes and mega-floods. Catastrophes are now helping to shape a new understanding of the world fit for a new millennium. Bearing in mind the vulnerability of scientific ideas, many contemporary so-called geological facts will yet be proved to be fiction. The planet we live on is, in many respects, still a mysterious place to us.

THE GEOLOGICAL TOUR GUIDE

Just as the Mediterranean provides insights into geology – and has been itself the site of many important observations and breakthroughs on how the natural world around us works – so geology can illuminate the human story of the region. The Mediterranean owes its shape, its architecture, its culture, its food, the basic realities of its life, to its geology. The rocks that geology has left on our doorstep are the building blocks on which past cities and empires have been built. Locked within the rocks are the raw ingredients for tools and trade on which the societies of the regions were built. Our understanding of the planet's restless climate can provide the backdrop to understanding the past environments in which our ancestors lived and how they lived. The realisation of the physical processes of Earth – earthquakes, volcanoes and cometary impacts – is a framework for charting the rise and fall of past civilisations.

Geology simply provides the rock-tinted spectacles to see an ancient world with fresh eyes. Learning how to read the rocks and interpret the landscape opens new insights into familiar places. It can take us beneath the skin of a region many of us feel we already know. A land of remarkable rocks and awe-inspiring natural wonders opens up to reveal a geological laboratory in the hidden depths below.

The first consideration is how the disparate lands of the Mediterranean got to where they are today, and where the water for the sea that connects them came from originally.

The Mediterranean, as depicted in a map of the world made from decorative limestone rock and displayed at the foot of the Vasco da Gama tower in Lisbon, Portugal.

CHAPTER 2

THE MEDITERRANEAN JIGSAW

To us, a sea may seem eternal as indeed a city may, a mountain range fixed till the end of time – for we, like all life on Earth, are short-lived carbon-based beings, whose timescale is biological, tuned to the way our brains and bodies work at optimum efficiency. Our planetary home, however, operates on geological time, and over the thousands of millions of years that Earth has had to play with, the jigsaw of lands and oceans on its surface has been made and unpicked countless times. For the ancient Greeks, the Mediterranean's scattered lands were the fallout of a conflict of cosmic forces between the Gods and the Giants when enormous rocks were hurled around by both sides: Athena killed one giant by hurling the whole of Sicily on him; Poseidon crushed another with a piece of the Greek island Cos, forming the island of Nisyros. Today, modern geology's explanation of those same lands may sound equally fantastic: they are the casualties of collisions between great moving slabs of Earth's crust. Much of Sicily is a fragment of Africa that was first grafted on to Spain and then ripped off, while lowly Cos was once a peak in a great mountain range that rivalled the Himalayas, and Nisyros is a melted blob of Africa. The geological story of how the Mediterranean came to be is one that is far more tempestuous and violent than any myth, and it starts, as any story does, at the beginning.

IN THE BEGINNING

Every culture at any time and at any place around the world has come to terms with its existence through, it seems, a particular creation story. There may be a single creator god, or a whole pantheon; a primeval void waiting to be filled, or a pre-existing world waiting to be shaped. Earth takes its place in the firmament along with that great blazing disc that travels every day over the land from horizon to horizon in a huge arc; with that mysterious white disc that waxes and wanes and sheds ethereal silvery light on its own travels; and with those twinkles of light in the night sky that may wander around or stay as fixed and unmoving in the universe as Earth itself. We know, from their Book of the Dead, that to the ancient Egyptians the land beneath their feet had originally arisen out of the floodwaters of a primordial ocean. From this watery chaos emerged the first solid matter – a primeval mound bearing the creator, the sun god Ra. Then, depending on your preferred version or sensibilities, Ra spat or masturbated to create his son Shu and his daughter Tefnet, who represent the properties of dryness and moisture. They then produced two 'children' of their own – the earth god Geb and the sky goddess Nut. From this union came the separation of the solid dryness of the earth from the watery expanse of the heavens.

Given that much of ancient Egypt was annually drowned by the flood-waters of the great Nile river, the idea of the land of the pharaohs emerging out of a vast waterworld is understandable, a compelling reconciliation of observation and speculation, and their speculation was partly right – in geological terms, Egypt did indeed rise from the waters of a great ocean. However, five thousand years on, and with the benefit of modern scientific thinking, we have a rather different take on how Earth came into being, and, thankfully, it didn't involve bodily fluids. While, as ever, we should remember the vulnerability of scientific theories, it's now not unreasonable to accept that everything started with that famous Big Bang.

According to this theory, the universe was born around fourteen billion years ago – in an explosion of unimaginable power that brought into being matter, energy, space and even time itself. Within a billionth of a second it was hundreds of millions of kilometres across and had temperatures of tens of trillions of degrees. In this fraction of a second, a seething soup of particles was created. At first, only electrons and quarks could stand the intense heat but, after just a millionth of a second, the maelstrom had cooled enough for some quarks to combine to form protons and neutrons. Electrons, neutrons and protons became a swirling mass of particles that began to clump together to create atoms, the basic building blocks of all substances. The earliest, and simplest, atoms were of hydrogen, which contained only one proton and one electron, but from about one second after the Big Bang, collisions between protons and neutrons started to make other atoms. Helium was next, its atoms having two electrons, two protons and two neutrons, followed by lithium which has three of each and beryllium which has four. This process was completed in three minutes.

Five hundred million years after the Big Bang that started it all, there were just those four chemical elements in the universe: hydrogen, helium, lithium and beryllium. Gravity began to cause these gas atoms to come together, and thirteen thousand million years ago they clumped together as the first galaxies; while further concentration of matter began to make the first stars. The production of heat and light in these new stars forced the four basic chemical elements to fuse into more and more complex arrangements of particles within atoms. Many of Earth's most common elements, such as oxygen, silicon and iron, were made in the reactors of these stars. The heaviest elements, such as lead, cannot be created in ordinary stars and had to wait for supernovae – the massive explosions of giant stars in their final death throes. These explosions also distributed the new elements throughout galaxies, where they were incorporated into new stars and planets. In one particular galaxy, the Milky Way, a clump of dust and gas began to condense about four and a half billion years ago and the solar system was born. It was out of this gassy stew of elements that first the sun, and then Earth and its rocky neighbours, formed.

Back in the early days of the solar system, Earth wasn't really Earth. It was a proto-Earth, a smaller world growing rapidly as it gobbled up rubble from the dust cloud swirling about the young sun. As the planet's fiery mass began to cool down, the swirling soup of chemical elements slowly settled out. The weightiest materials, mainly heavy elements like iron and nickel, sank to the bottom, forming a dense core. The materials that weren't as heavy stayed in the middle to form what is called the mantle. The same process can be seen in miniature when a glass of Guinness is poured, with the separation of the core – the black stout at the bottom of the glass – from the mantle, the overlying froth. On Earth, the very lightest materials, particularly silicon, floated to the surface to form a third layer – the crust.

But proto-Earth had a rival, another planet about a tenth of its mass, spinning around the sun in a dangerously similar orbit. About four and a half billion years ago, the two collided. The impact of this collision melted Earth's crust and obliterated the smaller planet, which some astronomers call Theia. Theia's iron core sank into Earth while its molten rock layers splashed out into space. Once in orbit, the splattered fragments of molten Theia were pulled together by gravity to form our moon. In astronomy circles, it's called the 'Big Splash'. It is the reason that the moon is made of lightweight rock, with iron making up only 8 per cent of the moon compared to 30 per cent of Earth.

Back on Earth, a thin crust was covering a searingly hot, molten interior. The early crust was little more than thin sheets of slag, formed by thickening lava piles spewed out from thousands of volcanoes. This congealed mat of mantle-like material was continually being pushed together, amalgamating, breaking up and rejoining, often being pushed back down into the hot interior. Volcanoes were pumping out gases like nitrogen, carbon dioxide and water vapour to form a primitive atmosphere, but little of this water vapour condensed to form surface water – it was quickly vaporised away by the intense heat. Instead, the water that would give our planet its first seas and oceans came from outer space. Earth's hardening skin was bombarded by comets and asteroids. The icy comets had a cooling effect, and water vapour, nitrogen and carbon dioxide rained down through the early atmosphere. Downpours of hot acid rain lasted for millions of years before Earth's surface had cooled enough to let water condense rather than being vaporised away. Pools of warm water grew to form lakes, then seas and ultimately, around 4,000 million years ago, the first oceans. Water had begun to flood the surface of much of the planet.

The early oceans were the cauldrons of life. The first organic molecules needed for life evolved in the primordial oceanic soup, and by 3,000 million years ago, micro-bacteria were using the sun's intense energy to begin to convert the masses of carbon dioxide in the early atmosphere into carbohydrates, kick-starting the process of photosynthesis. A by-product of this reaction was oxygen, a highly active gas that readily combines with other elements or substances to form oxides. Over millions of years this waste product of photosynthesis began to oxidise the rocks of Earth's surface. Earth's volcanic crust began to be eaten away by the acid rain, and the clay-rich residue washed down streams and rivers to be deposited into lakes, rivers, deltas and beaches, or swept offshore to settle on the ocean bed. Sediments began to build up and be transformed into rock.

The planet was beginning to recycle itself. As it did so, the light minerals such as silica (quartz), which is virtually immune from chemical breakdown, and aluminium became increasingly concentrated in the upper layers of the crust. Over time, this buoyant 'froth' would clump together to form the first continents. The earliest continental crusts are around 3,800 million years old, but most are about 2,000 million years old. These most ancient parts of the planet lie in the great interiors of the continents. By comparison, the oldest continental rocks around the Mediterranean are far younger. The rocks of eastern Egypt and western Arabia are among the region's oldest, dating from 1,600 to 600 million years old, but it is only at the end of this period that the backbone of other parts of the Mediterranean begins to form. Six hundred million years ago, sandy and silty sediments were being laid down on an older basement in north-eastern Greece and north-western Turkey, while in western Spain thick deposits of mud were forming around 550 million years ago. Young as they may be by Earth's standards, these are the Mediterranean's ancient heartlands.

MOVING PARTS

The ancient Greeks believed that their sacred island of Delos had wandered through primordial seas until Leto, made pregnant by Zeus, gave birth to Apollo there, at which point four diamond pillars rose up and anchored the floating island to the Cyclades. In fact, geologists now know that not only Delos but the entire surface of the planet is on the move. That's because Earth's outer shell is broken into a number of rigid slabs, called plates, and, like a great jigsaw puzzle, these plates are endlessly jostling with each other. Today there are seven huge plates – North America, South America,

Caribbean, Eurasia (Europe and Asia), Africa, Pacific and Antarctica – all complex mosaics of continental land masses and ocean basins.

The popular notion of the Earth's plates is of a rigid crust of lightweight rock riding atop a semi-molten sea of denser mantle material, but in fact it is not just Earth's crust that is moving, it is also a thick chunk of the mantle. From the ground surface to a depth of about 100 kilometres, the temperatures are not high enough to melt rock; geologists call the cold, rigid upper area the 'lithosphere'. Deeper than 100 kilometres, the temperatures are hot enough to melt rock, but the pressure of overlying strata is now high enough to keep rock so tightly packed that it is prevented from melting. This warm, compact area is called the 'asthenosphere'. However, at around the 100-kilometre depth, the temperature and pressure are just right for a small amount of the mantle rock to be able to melt. The result is the formation of a thin weak layer of semi-molten rock – the low-velocity zone – that acts as a lubricant to the rigid mass of colder rock above, allowing it to slide over the closer packed hotter rocks below. Rather than just two layers – lighter crust and denser mantle – Earth's outer shell could be seen as a layered sponge cake with icing! The hard icing is Earth's crust, and the top layer of sponge is denser mantle rocks. Together these constitute the lithosphere. A thin coating of jam separates the top layer of mantle sponge from the bottom layer of sponge. The jam represents the weak low-velocity zone, while the bottom sponge is the warm but tightly packed mantle rocks of the asthenosphere. If the cake is squashed, then the layers of mantle sponge slide along the jam rather than the icing crust.

What drives the Earth's hard shell – its lithosphere – to move is what is happening deeper down inside the planet. Just as the ancient Egyptians believed that Shu had raised Nut, goddess of the sky, away from Geb, god of the earth, so modern geologists believe our planet slowly differentiated itself by heat loss. The atmosphere contains the lightest elements, and, since these can retain only minute amounts of heat, they languish at the coldest limits of our world. By contrast, the densest elements reside in the planet's iron-nickel core where, because of their ability to retain heat, temperatures exceed 4,000°C. The core is the critical heat source for the planet. The transfer of radioactive heat from the planet's nuclear core to the base of the lithosphere is the fundamental process which operates within Earth – it is the engine that drives the plates.

That engine behaves much like a giant lava lamp. In a lava lamp, heat from the light bulb in the base of the lamp causes the waxy substance to expand, making it lighter than the viscous liquid above it and allowing it to rise. As the wax plume rises, it cools and slowly reverts to being more dense than the

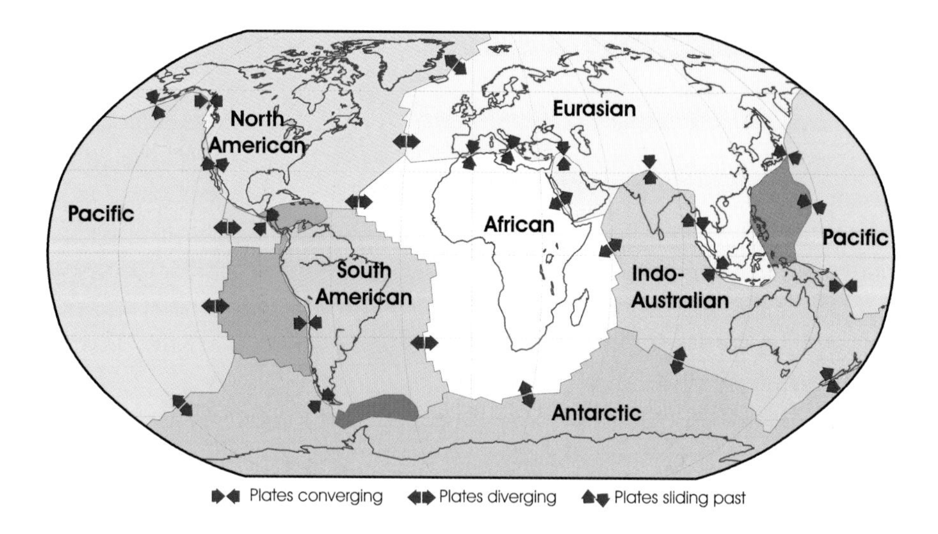

surrounding fluid, so it falls back down again, only to be reheated . . . and the cycle repeats endlessly. In just the same way, heat lost by the Earth's light bulb – the radioactive iron-nickel core – passes outward to the mantle, warming the lowest layers and making them less dense. Blobs of the strong but malleable mantle rock creep slowly upwards until they reach the base of the crust. Some light material sticks to the underside while most of it cools and descends again into the mantle abyss. It is this unrelenting convection of heat from the core to the crust that forces the rigid outer layer to move.

So, the planet's plates are moving on a thin layer of 'jam' – that is, semi-molten rock – stirred by the slow swirling motions of hot plumes below. This is plate tectonics. Every part of Earth's surface is on the move. In some places, the plates just slide past each other, without any crust being gained or destroyed. In others, rising plumes blister a plate, it breaks apart, and hot mantle rocks seep out through the split and feed a widening new ocean floor. And if some parts of Earth's shell are spreading apart, others must be compressed together. When a plate of dense ocean floor is squeezed against a plate of light continental rock there is invariably only one winner. The dense, thin ocean floor generally gets pushed down below the buoyant continental mass and its rocks return to their mantle womb, carried to the depths of a few hundred kilometres by a one-way escalator called a subduction zone. As the cold slab of ocean rock sinks into the hot, plastic interior it drags down the remaining ocean floor still at the surface, until the entire ocean is pulled closed. With the ocean shut, previously far-flung continents now collide; both are equally buoyant and, like two wrestlers each unwilling to succumb to the other, the collision results in geological violence. Here the crust is contorted, stacked and thickened into great mountain chains.

MAKING UP, BREAKING UP

Ever since the first continents formed, they have been a restless congregation of moving parts – shifting, separating and colliding plates. Earth's plate motions seem to switch regularly from dispersing its continental fragments about the globe to clustering them back together into a single landmass – a supercontinent – surrounded by a global ocean. Five hundred million years ago, far-flung continental landmasses were converging. One fleet of landmasses was moving northwards from their earlier moorings around the south pole. These would later become the lands of South America, Africa, Australia, India and Antarctica, but back then they were a motley crew of fused bits and bobs. In the northern hemisphere, two other enormous landmasses were approaching. North America and Greenland, with northern Britain in tow, were separated from northern Europe, the Baltic countries and northern Asia by a wide ocean. That ocean was given the name Iapetus, in Greek mythology a fruit of the union between the Earth goddess (Gaia) and the god of the Sky (Uranus) – modern terminology happily marrying science with echoes of the ancient mythologies it has superseded.

Separated by a shrinking ocean, the landmasses slowly closed in on each other. By 350 million years ago only a narrow tropical seaway separated the northern continents (Laurasia) from the southern (Gondwanaland). A hundred million years later and this body of water was sealed shut. The multi-continent pile-up had thrown up great chains of coastal and inland mountain chains, Alpine, possibly even Himalayan, in stature. Now the world's continents were fused into a single continental amalgam called Pangea, an enormous landmass that straddled the equator in the middle of the world. Around the edge of Pangea were new oceans that had spread open as the ancient ones had shrunk closed. To the west, an extensive sea body, Panthalassa, would become the future Pacific Ocean. To the east, wedged between the frayed continental margins of Africa and of Europe and Asia, was a smaller equatorial ocean, a sister of Iapetus – Tethys.

While Panthalassa was all-embracing, Tethys was still about the size of today's Indian Ocean. Two hundred and fifty million years ago its waters stretched from Spain in the west to the Asian archipelago of Vietnam to Borneo in the east, bordered by vast continental shelves and shallow continental seas basking in a warm, wet tropical climate. In the western Tethys, the inland plains and hills of central Europe marked its sun-drenched northern shores while to the south, sub-tropical seas lapped against shores fringed by coral reefs from Tunisia to Lebanon. Between those opposing shores were vast shallow continental plains which, with the ups and downs of world sea

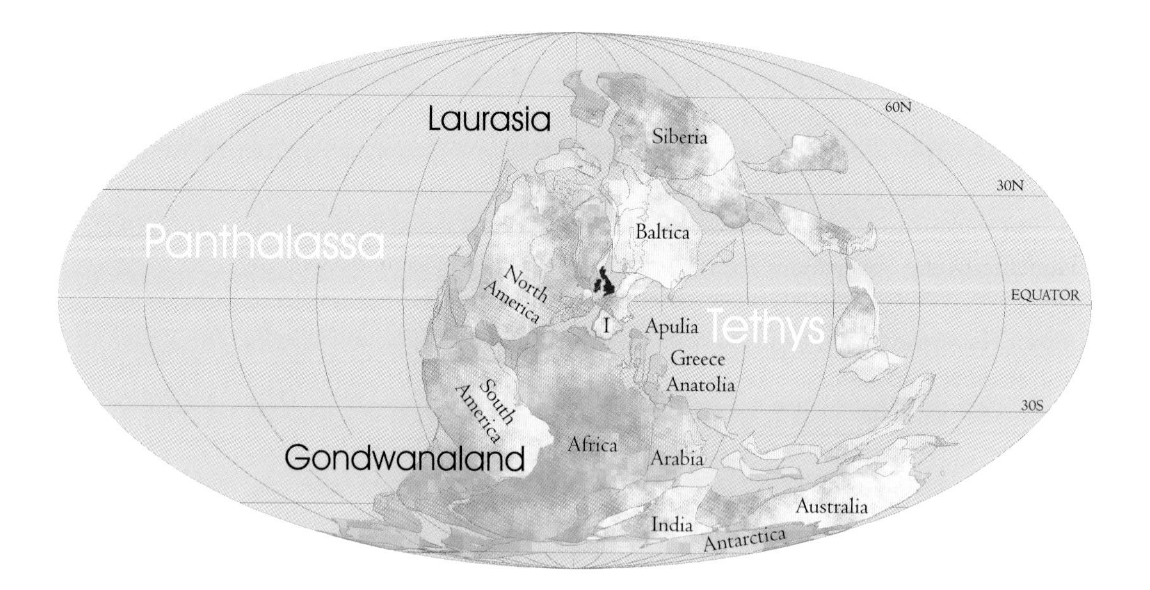

The fotsam and jetsam that made up the great equatorial landmass of Pangea two hundred and fifty million years ago was an amalgam of a set of southern continents (Gondwanaland) that became welded onto a set of northern continents (Laurasia). Wedged between what would later become the landmasses of Europe and Asia on one side and Africa and Arabia on the other side was the Tethys ocean.

level, switched countless times between emergence and submergence. Eventually, from the northern continental shelf would rise the eastern Alps, the Balkan states, the Danube basin, the Carpathian and Caucasus mountain chains, and Iran. From the southern shelf would emerge Italy, Croatia, Greece, and Turkey.

Two hundred million years ago, the great sutured landmass of Pangea began again to tear its old wounds apart and the immense Tethys ocean began to close. The first cracks on Pangea appeared between eastern North America and West Africa, when what is now New York sat next door to Mauritania; today the two are separated by over 5,000 kilometres. A string of volcanoes that stretched from the Carolinas to Nova Scotia, and from Agadir to Rabat, erupted as the continent began to be pulled apart. As the continent split, a rift valley, much like the lake-strewn Rift Valley of East Africa, began to develop. Gradually, waters from the Tethys Ocean spilled into the deepening trough through a narrow seaway at Gibraltar. The Atlantic Ocean had been born.

Over the next few tens of millions of years, the Atlantic Ocean simply unzipped itself slowly southwards and northwards. By 144 million years ago, a jagged split began to slowly zig-zag its way southwards into the southern continents, forcing apart South America from Africa. By 110 million years ago, a similar split crept northwards to allow the North Atlantic to open, shunting Europe eastward as North America spread to the west. In the south, the breaking away of South America forced Africa to swing northwards, disintegrating as it did so. India and Madagascar jumped ship early on, but the rest of the continent, with Arabia in tow, was on a collision course with Europe and Asia.

WATERWORLD

By 144 million years ago, the start of the Cretaceous period, the dispersing fragments of Pangea had created a network of new seas and small oceans. The newly formed ocean floors were swelled upwards with the hot and temporarily buoyant mantle rocks pouring out of the great splits. With less space in the global ocean, the seas began to flood the land, drowning almost two-thirds of what today is dry land. Onshore, the rising waters ensured that plants and animals, including the dinosaurs, were confined to ever smaller and more isolated uplands, but in the oceans, life had never had it so good.

The equatorial waterways of the Tethys were teeming with life. The most abundant life in the oceans was the simplest – countless trillion trillions of unicellular and multi-cellular organisms called protocists. Far more primitive than plants or animals, these organisms first began to evolve about 2,000 million years ago when photosynthesis had made the oxygen-rich environment poisonous to the very bacteria that had created it in the first place. Some bacteria adapted to use the ever more plentiful oxygen, but others didn't. As more and more bacteria became oxygen-dependent, or aerobic, some anaerobic ones survived by invading their new neighbours. The result of these invasions was more complex single-celled organisms in which the cells developed internal units that had different roles and responsibilities – respiration, photosynthesis and movement. For a thousand million years, these organisms slowly developed until something happened that led to the explosion of the first multi-cellular organisms.

Between 750 and 550 million years ago, the planet was racked by a series of severe global ice ages that covered it in ice from the tropics to the poles. Earth, it seems, became a giant snowball. The dramatic switch of climate forced major changes in world sea levels and the chemistry of the oceans – setting the scene for an explosion of multi-cellular life which, when the global ice sheets disappeared, was ready to evolve. It is now that the first soft-bodied animals appear in the ocean: 550 million years ago Earth had just entered the Palaeozoic – the era of 'ancient life' – and the next few hundred million years would see the appearance of a large number of organisms with shells and hard parts.

The first fishes appear 500 million years ago, initially as soft, toothless 'slimefish' and later as jawed fishes with armour plates that would develop into scales. By the Cretaceous, some of these early fishes had developed into sharks and rays that bear a striking resemblance to their modern counterparts, but they were vastly outnumbered by bony fish – salmon and trout, cod and flounder, herring and mackerel all had their ancestry in these Cretaceous waters. The ancestors of today's squid and octopus also first appeared about 500 million years ago and spawned a vast number of cephalopod species. Some of these, like the convoluted ammonites, were completely extinct by the end of the Cretaceous; today the last remaining cephalopod living in a shell is the straight-coiled nautilus – a distant relative of the ammonite that thrives in the equatorial waters of the Indian and Pacific oceans. Drawn by the rich source of marine food, some land reptiles adapted to life in the oceans. Shark- or tuna-like reptiles – Ichthyosaurs – and larger creatures resembling the short-tailed, long-necked Loch Ness Monster of popular imagination – Plesiosaurs – prowled the seas, before both became extinct.

However, despite the richness of life in the Cretaceous waterworld, by far the most important were those small fry – the protocists. Many are single-celled micro-organisms, such as radiolarians and foraminifers, while others form plant-like colonies of red and green algae, and others still are more complex multi-cellular organisms such as brown algae (kelp and seaweed). Protocists were (and still are) the main contributors to the food cycle that sustains the life in the oceans. Many of them are also the prime contributors to marine sediments. The great chalklands of the region are made from the skeletal remains of microscopic single-celled algal protocists called coccoliths, while the silica-rich hard parts of the radiolarians formed oozing matter that solidified into nodules of chert or flint. The hard calcareous parts of foraminifers built up in huge masses to form the extensive lime-mud sea beds that would be transformed into limestones. The soft parts of protocists were also responsible for the enormous quantities of organic remains that became buried in the sea bed and later matured into the vast oil and gas reserves of the region.

■ WORLDS IN COLLISION

By 140 million years the great African plate, which had been sliding past the Eurasian plate, began to swing northwards and crunch directly into it. Europe remained almost motionless as the intervening ocean floor began to be destroyed and the great landmasses of Africa edged ever closer. As the African plate twisted round, fragments of its northern margin were ripped off and dispatched northwards. They began to arrive along the European shores of Tethys around 100 million years ago. Gradually an armada of geo-

Tilted and folded limestone strata in northern Turkey are the result of the muddy sea bed of the Tethys ocean being crumpled and pushed up into mountain ranges when fragments of the African continent ploughed into the European landmass a hundred million years ago.

logical flotsam and jetsam began to ram into the northern landmass. The first to dock were the slivers of crust that would make up the northern lands of Greece and Turkey, the Caucasus region and Iran. Most had been the floors of shallow limestone seas, but as they ploughed into the hard ancient basement cores of areas the sea bed sediments were pushed up to form a rim of mountains. Long subduction zones formed along the edge of Eurasia, consuming the old, cold ocean floor of Tethys.

Further south from the main collision front, the shallow continental shelf of Arabia buckled like a crumpled carpet. From 90 million years ago, range after range of folds were thrown up – forming an arc of uplands that today stretches from the Palmyride mountains of Syria, south into the coastal peaks of Lebanon, the hilly uplands of Israel and west into the mountainous Sinai peninsula – but just as some parts of the Arabian shelf were buckled up into broad swells, others were pressed down to form deep troughs. In these troughs, from Libya to Oman, thick piles of organic-rich sediments built up. As the Arabian block continued to crumple as it neared the collision zone with Europe, these sediments were pushed ever deeper down and buried, the

Flat-lying rock strata of the Arabian plate became gently warped, buckled and bent as it crunched into Europe.

resulting heat and pressure cooking the organic material into oil and gas. Later, about 30 million years ago, when the main solid heart of the Arabian landmass would crunch into Europe, renewed spasms of folding and mountain building would raise these deep offshore basins up to form the rich shallow oilfields of the Middle East.

Cyprus's northern Kyrenia mountains, a long, jagged blade of peaks, started out as soft sediments in warm African seas on the shallow shelf of an archipelago of large islands, much like the Caribbean is today – low-lying patches of land fringed by sandy beaches and coral reefs, while deeper offshore thick piles of mud accumulated. By 80 million years ago, however,

the coral islands of Kyrenia were being plastered against the mountainous rim of the great Anatolian plateau of Turkey, and around the same time, Cyprus's southern half was being formed several hundred miles to the south. What are today the 2,000-metre-high peaks of the Troodos mountains were once molten lavas spewing out of a great split in the ocean floor. The split was creating new ocean crust, but to the north the Tethyan ocean floor was being destroyed faster than it was being made, so gradually the Troodos rocks too were dragged northwards to be subducted. Perhaps because the ocean rocks were still warm and slightly buoyant, when the Troodos sea floor collided with Eurasia around 25 million years ago this sliver of ocean crust didn't disappear down the hatch. Instead, a 100-kilometre-long lump of the ocean floor was unceremoniously thrust up on to the messy Turkish margin. With Troodos now firmly wedded to Kyrenia, the rocks of Cyprus were geologically united, but they were also still firmly attached to Turkey.

From around 100 million years ago, the Aegean region was witnessing a gigantic pile-up as a train of enormous continental blocks embedded themselves into the ancient continental heartlands of Thrace and north-eastern Turkey. Pelagonia, a shallow limestone sea floor, arrived to form the rocks of much of northern and central Greece. Behind it came another shallow limestone sea, the Pindos. By 35 million years ago this had been squeezed shut, and it thrust up and over the rocks of Pelagonia. One by one, successive detached fragments of the African continental shelf ploughed into the back of the Pindos and Pelagonian block. Each impact increased the squeezing and crumpling of the ones in front and the whole area became a mountain chain, the Hellenides.

Similar collisions were ruffling up the edge of Europe. For most of the history of Tethys, Italy had been part of a vast low-lying landmass on the northern edge of Africa. This mass, stuck out as an enormous rigid prong, was called Apulia and included, with mainland Italy, the lands of the Adriatic – Croatia, Albania and western Greece. Around 65 million years ago this prong sheared off, and by 50 million years ago it was ramming into Eurasia, throwing up the Swiss-Italian Alps and the Croatian Dinarides. Further west other fragments of Africa embedded themselves on Spain. Trapped in a vice between the might of Africa and Europe, the ancient Iberian landmass anchored itself on to Europe, the suture between the two being the Pyrenean mountain chain that today divides France and Spain.

About 70 million years after the first collisions, the geological shrapnel from the breakup of the African continental shelf had transformed the southern edge of Europe into a great chain of mountains – the Spanish Betics, the Pyrenees, the Alps, the Dinarides and the Hellenides, and the Taurides.

SUNKEN KINGDOMS

Then, about 20 million years ago, many of these mountain ranges began to collapse. Previously the great peaks had been propped up by the battering ram of new continental land masses crashing into Europe, but most of these wandering African fragments were now attached. Below, the Tethys ocean floor descending deep beneath Europe was cold and dense, and slowly the slab of oceanic rock began to sink down into the mantle. As it sank it tugged away at the continental crust above it, and mountains that had been forged through enormous squeezing now began to be pulled apart. Great cracks appeared at the heart of whole mountain ranges as the huge forces needed to keep them held up began to wane. Around the Mediterranean, they began to slide back down.

Along the southern edge of Turkey, the Tauride mountains began to be ripped apart. The subduction zone lay to the south of Cyprus and as the African plate sank it pulled away at the southern margin of Anatolia. The thickened crust began to be dismembered and, as great splits appeared, molten magmas erupted out, forming the spectacular volcanoes of Cappadocia. Over a distance of 300 kilometres, central Anatolia was peppered by volcanoes. The eruptions began around 14 million years ago, almost dying out by three million years ago and becoming all but extinct by one million years ago, but the main consequence of the fragmentation of the Tauride range was that the newly wed Kyrenia and Troodos blocks broke free from the Turkish landmass, drifting south to form Cyprus.

In the Aegean, great wrenches appeared in the Hellenide mountains. In the north the arthritic fingers of the Chalkidiki peninsula were ripped away from the cloud-covered crown of Greece – Mount Olympus. In central Greece huge faults broke up the towering rocky spine and left large areas like Evvia and the Peloponnese hanging by a thread to the mainland. Western Turkey too was being slowly dismembered, the mountainous terrain of the Menderes massif being splayed open to allow great river valleys to form. Across the region, rocky debris eroded from the mountain peaks was building up as huge piles of sediment. In central Greece, these piles would eventually give rise to one of the world's natural wonders, the awe-inspiring rock towers of Meteora. Rising like a giant rocky forest out of the Thessaly plain are more than twenty lofty pinnacles, most with medieval monasteries dizzily perched on top. The sheer cliffs, most rising to more than four hundred metres, are made entirely of gravels and pebbles washed down from the erosion of the nearby Pindos mountains and then transformed into enormous natural sculptures by incising rivers.

The greatest stretching, and so the most severe collapse, occurred in the heart of the ancient Aegean mountain range. Here, mountain peaks were lowered to such an extent that they disappeared below sea level. They lie drowned in the seas north of Crete. Scattered around this are relict peaks that haven't yet been pulled down beneath the waves; instead they jut out as the Greek islands – the Cyclades, the Sporades and the Dodecanese. Beyond this are the contorted gulfs, bays and inlets of the mainland coast of Turkey and Greece where the fingers of the Aegean sea remorselessly invade a landmass that is still sinking and which is destined for a watery grave over the next few million years.

Even more dramatic were the changes further west. Twenty million years ago the sinking African slab beneath Spain similarly began to tear apart the continental landmass above. In southern Spain, a wide mountain range that stretched from the Atlas Mountains to the Spanish Betics imploded, its central peaks collapsing down to be drowned in the narrow seaway that

Towering above Palermo in Sicily are the limestone cliffs of Monte Pellegrino.

The whitewashed town of Oia clings to the caldera cliffs of the volcanic island of Santorini in the Aegean Sea. The island's spectacular red rocks are formed from ashes and lava flows that have erupted for the last million years or so, though the sheer walls here were exposed by the volcano's catastrophic blast over three and a half thousand years ago.

separates Spain and north Africa. Further north, violent volcanoes erupted from Spain's Costa Brava through French Provence to the western Alps. Later, the slivers of Africa which had a few tens of millions of years earlier been plastered on to eastern Spain now broke off and were dragged eastwards. Gradually further splits broke this enormous block of crust into chunks – first detaching the Balearic islands of Majorca and Menorca, later the islands of Corsica and Sardinia, and finally sending off the Calabrian toe of southern Italy and parts of northern Sicily. It was this eastward drift of this flotilla of lands that crumbled up the western edge of Apulia to form the mountain spine of the Italian Apennines.

The final ignominy of the once great mountain ranges was that in some places the continental crust had become so pulled apart that mantle rocks from below were able to break through and burst out. Around seven million years ago three great splits appeared in the ocean floor to the west of Italy creating the Tyrrhenian Sea and gradually these crept towards the mainland. Around five million years ago the southern split reached southern Italy, and the volcanoes of the Aeolian islands appeared. Later, around one million years

ago, a northern split also allowed volcanic outpourings in Rome's backyard, and a central split formed the volcanoes of the Naples region. The northern volcanic centres have all but switched off, but the Neapolitan and Aeolian ones remain active. At the same time, volcanoes were erupting in the southern Aegean, producing the volcanic islands of Milos, Nisyros and, most famously, Santorini.

MEDITERRANEAN CRISIS

For over a hundred million years, the enormous Tethys ocean had been slowly strangled by an ever-tightening noose of lands, but the most dramatic events were saved for its final geological moments. Ten million years ago, the final connections to the Indian ocean were severed as the main mass of Arabia finally docked with Europe. At the same time, at the opposite end of the seaway, the marine passages in southern Spain and northern Morocco that linked the shrinking sea with the Atlantic began to be squeezed shut. Around six million years ago, as world sea levels were dropping by around sixty metres as the growth of the Antarctic ice sheet drew moisture, a final spasm of land uplift raised the shallow Straits of Gibraltar forming a land-bridge between Morocco and Spain that allowed mammals to cross. Cut off from the fresher Atlantic waters, the enclosed sea lost most of its water through evaporation. This was in spite of the inflow of fresh water from innumerable rivers and the drainage from the Black Sea where the great Russian rivers were delivering huge volumes of water. The water level dropped by more than one thousand metres within a thousand years – a geological instant.

The sea was transformed into two great salty lakes, one in the east and one in the west, separated by the desert landbridge that connected Sicily, Malta and Tunisia. Tethys became an evaporating bath-tub that filled up with thousands of metres of salt. The amount of salt that collected on the sea floor is estimated to have been in the order of several hundred thousand cubic miles – more than thirty times the amount of salt contained in the present waters of the basin. At the times when world sea levels rose, the basin was replenished by a gargantuan waterfall that cascaded over the Gibraltar sill, only for the basin to dry out again when global ocean levels fell. Great waterfalls developed at the mouths of the main rivers as they were unable to cut down as fast as the water levels plummeted. Over time, however, the largest rivers did start to adjust to the far lower sea levels, cutting enormous rock gorges. For instance, the most formidable river, the Nile, cut a canyon deep into the African continent that is three times longer than the Grand Canyon and

with a similar width and depth. Other canyons of the same age are known to exist beneath the Rhône, the Ebro and the Po rivers and many of the deep ravines and gorges throughout the Mediterranean were created during this catastrophic dry spell.

About five million years ago, the great splits that were breaking up the Spanish Betics and Moroccan Rif mountains breached the landbridge at Gibraltar, and the Atlantic waters spilled back into the salt basin, flooding the bones of the ancient Tethys. Within a few thousand years, the water levels returned to match those of the global oceans. The Tethys Ocean had become the Mediterranean Sea.

Today, without the narrow gateway at Gibraltar that connects it to the Atlantic Ocean, the Mediterranean would quickly dry up. At Gibraltar, the Atlantic and the Mediterranean trade salt. Atlantic surface currents funnel cold, less salty water in, while deeper, denser salty brine is pumped out. Even with this life-saving transfusion, the world's saltiest open sea is getting ever saltier. Rivers feeding freshwater to the Mediterranean are unable to keep pace. Even though huge amounts of it flow into the sea, predominantly cold water from Russian rivers draining into the Black Sea or the Nile waters bringing tropical rains from the African interior, they still make up only about one-third of the water lost by evaporation. If the Gibraltar Straits were dammed now, the sea would again dry up in about a thousand years. As the plate motions continue to carry Africa northwards into southern Europe, it is only a matter of time before the gateway closes again – and when it does the Mediterranean Sea, like the Tethys Ocean before it, will be history.

Opposite: The Mediterranean's rocky coastline are dissected by hundreds of deep gorges and ravines, such as this one slicing down through limestone cliffs at Arvi in southern Crete. Many of these rock canyons began to form around six million years ago, when a great drying out of the sea caused water levels to plummet by hundreds of metres and rivers were forced to follow suit.

THE MODERN MEDITERRANEAN

This land-locked backwater of the world's oceans is in some respects an inconsequential part of the planet. Yet the roots of the western world sprang up around the shores of the Mediterranean, and for the earliest peoples the lands that its waters lapped constituted the known world. Successive peoples, powers and empires would have their own name for it, though often it was just 'The Sea'. The early civilisations like the Egyptians and Mesopotamians called it the Upper Sea, lying as it did on the edge of their great territories, but for those living on its shores, its open waters gave a unity to the disparate lands and peoples. By the first millennium BC the Phoenicians began calling it the 'Great Sea', and they should know since it was on the back of their ability to navigate it that they became the first Mediterranean-wide civilisation. To the ancient Greeks, it was the 'Sea over by Us' and, like the Phoenicians before them, they built their empire by embracing its great expanses (as indeed the Venetians would do two millennia later). Imperial Rome simply claimed it as 'Mare Nostrum' – 'Our Sea' – reinforcing their view that Rome was at the heart of the inhabited world, but it wasn't until the sixth century AD that the term 'Mediterranean Sea' itself emerged in the encyclopedic writings of Isidorus of Seville. According to him, 'The Great Sea [Mare Magnum] flows from the ocean in the west; it faces south and reaches north. It is called "great" because other seas pale in comparison; it is called the Mediterranean because it washes against the surrounding land [mediam terram] all the way to the east, dividing Europe, Africa, and Asia.'

How far inland of these shores does the Mediterranean world extend? What do we mean by 'Mediterranean'? To the travellers who flock to the region today, it is the climate – hot dry summers and mild rainy winters – that defines the Mediterranean for them. Strictly speaking, in climate terms Mediterranean lands are those that receive between 250 mm and 500 mm of rainfall per year. The resulting belt of modest rainfall takes in much of eastern Spain, Mediterranean North Africa, southern Sicily, eastern Greece and interior Turkey and the Levant coast of the Middle East; omitted are southern France, Corsica, Sardinia and the whole of Italy, and much of coastal Greece and Turkey. Rather than in arbitrary rainfall figures, a more obvious expression of the limits of the Mediterranean climate can be found in the vegetation. Commonly used indicators are the northern extent of the olive and the southern extent of Mediterranean vegetation. This delimits a great swathe of southernmost Europe and northernmost Africa, though areas like the interiors of Italy, Greece and Turkey are excluded on account of being too cold or dry for olive growth. For some, the best expression of the demanding climate is found in its human imprint, in a sense of shared common lifestyle and a common temperament of the people of the region

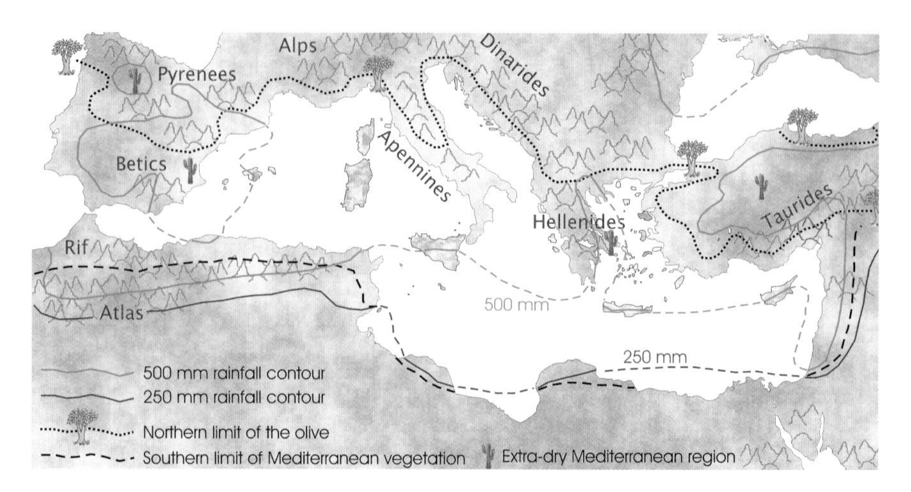

Physical definitions of the Mediterranean.

characterised by the traditional midday siesta or the sense of mañana and a laid-back approach to life.

In a way, all of these physical attributes of the extent of the shores or the nature of the climate are predicated by geology. The template of the Mediterranean Sea was set by the great movements of the African, Arabian and European plates and the architecture of its landscapes was sculpted by local idiosyncrasies of geological collision and collapse. The mountains enveloping, strangling, barricading and partitioning the Mediterranean area are the flesh and bones of the ancestral Tethys, and define the edge of those lands that have a shared experience of that ancient ocean and its demise. This is what the Mediterranean means to a geologist – the lands from the Straits of Gibraltar to the Jordan hills, and from the peaks of the Alps to the desert fringes of the Sahara. All have been intimately connected with an ocean that has long disappeared, but whose physical vestiges remain all around.

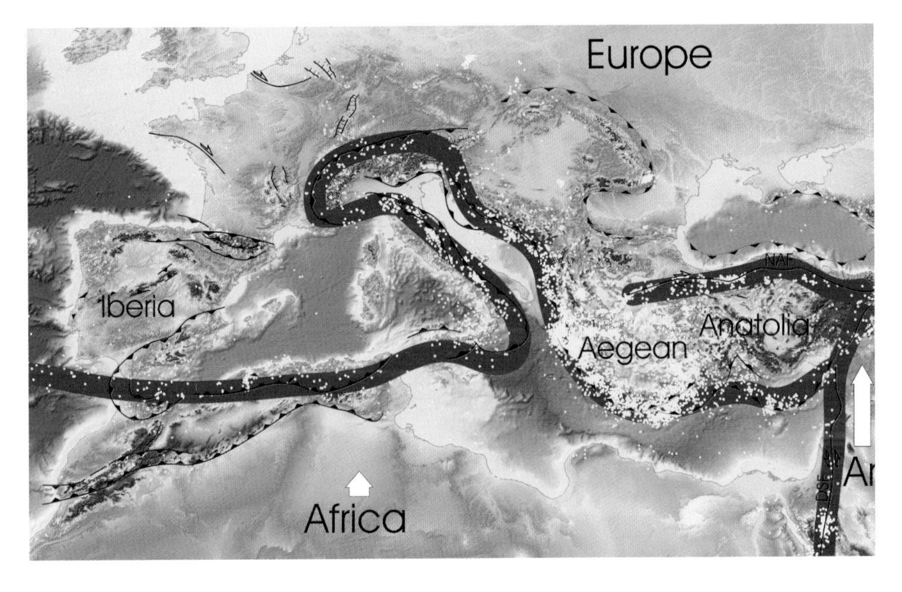

The present-day geological setting of the Mediterranean, with the main plate boundary zone depicted in red, mainly following the belt of modern earthquakes (yellow) and major faults (black lines). The northward motions of the African and Arabian (Ar) plates towards Europe are indicated by the white arrows.

Some of those ancient traces can be seen in the political divisions of the modern Mediterranean. Buckled ancient sea beds act as the barriers between countries. The Pyrenees, which once separated Iberia from Eurasia, now keep Spain distinct from France. The winding trace of the Alps wraps its way around the north of Italy, isolating it from the rest of Europe. In the east, the Tauride highlands separate the great Anatolian interior from Syria – effectively Europe from Arabia. In the south-east corner, the coastal crags of the Moroccan Rif on one side and the high Spanish Betics on the other seem almost like the natural ramparts that keep Europe and Africa apart.

It is the collapse of many of the mountain chains that has bequeathed the Mediterranean another legacy – its seas. The Mediterranean is not just one expanse of open water but rather a mosaic of interconnected basins that have grown out of the submerged ruins of the foundering continents. Between Morocco and Spain is the Alboran; further west is the Balearic, between the Balearic islands and the coast of Spain; and the Ligurian, between the Balearics and Corsica. Then there is the triangular Tyrrhenian, bounded by the northern coast of Sicily, the eastern coast of Sardinia and Corsica and the western coast of Italy. Between Italy and Croatia and Albania there lies the Adriatic. Between Sicily and western Greece there is the Ionian; and between the lands of Greece and Turkey there is the island-studded Aegean Sea. Within these marine environments there are further compartments – islands, gulfs and straits, most of which are again defined by geological inheritance.

The Tethys Ocean had been constantly changing since it formed, and now that it has become the Mediterranean it is still changing. Arabia, Africa and Europe are continuing to squeeze the lands of the Mediterranean in an enormous vice, but perhaps the most obvious legacy of Tethys lies in the web of splits and fracture lines that criss-crosses the region. These represent the modern boundaries of the mosaic of plates and crustal blocks that gave rise to the Mediterranean jigsaw. Where they are determines where the main areas of geological violence are in today's Mediterranean – the belts of volcanoes and earthquakes.

In the east, they lie hidden off the coral-draped shores of the Red Sea, and splinter up through the Gulf of Aqaba to pass along the Dead Sea and the Sea of Galilee. Here this great split in the continent, called the Dead Sea Fault, carves through the heart of the Holy Lands, ripping through places that have seen more than their fair share of violence. A strand peels off to pass below Beirut, but the main crack follows Lebanon's former war-torn Bekka valley to kink eastward through the mountains south of Damascus, then heads north again across the plains of Syria. All along this line, the Arabian side of the divide is moving northwards, as it tugs gently away from Africa. From biblical times through the Crusades and up to the present troubles, this has been a divided, fractured land. The Middle East tension that is all too apparent at the surface mimics the subterranean world.

Further north the plate boundary meets a great fracture line that links the subduction zone south of Cyprus to the volcanoes of eastern Turkey; beyond are the mountains of the Caucasus and Iran. From Mount Ararat – the alleged site of Noah's Ark – a great fault line heads westward, skirting along the northern coast of Turkey towards Istanbul then heading offshore to die

Once a war-torn battleground, the Bekka Valley in Lebanon defines a persistent line of geological violence, as a great fault line rips its way up through Israel towards Syria and Turkey, marking the western edge of the Arabian plate.

out in the Aegean Sea. Sliding along this fault, the Anatolian interior is moving slowly westward relative to a stable Europe to the north. It is spreading into the Aegean, but the southern islands of Aegean are accelerating away from it, riding roughshod over the African sea floor. That sea floor is diving beneath Crete at an angle of 35 degrees, sinking ever deeper and pushing the island up as it does so. We can hear and feel the effect every time Africa descends a little beneath Crete – these are the frequent earthquakes on the island. From Santorini, the edge of the African plate is now 240 kilometres south, or 160 kilometres directly below.

The Cretan subduction zone swings round the edge of the Peloponnese before brushing past the popular Ionian islands of Zakynthos and

Previous pages: Mount Etna, the largest continental volcano in Europe and by far the most active, towers over eastern Sicily and dominates the lives of over two million people who depend on its fertile slopes for countless vines, olive trees and orange groves. Despite its enormous size, the bulk of this giant volcanic edifice formed in the last half a million years.

Cephalonia and then Corfu. North of here the plate boundary subduction stops and instead the rigid Adriatic crumples into Albania and Croatia, forming a mountainous belt of violence that has seen frequent destructive earthquakes. Another plate boundary runs south along Italy's Adriatic coast, from the Po delta down past the Gargano peninsula and through the Apulian 'heel' of Italy, swinging northward to pass through Catania in eastern Sicily. The great Mount Etna springs up here, its magma rising up the great fractures in the crust as the African sea floor pushes below it. The plate boundary continues westward, swinging offshore to pass by the sunken island of Ferdinandea and then comes onshore beneath ancient Carthage, close to modern Tunis. From there it skirts along the north African coast, eventually passing out through the Straits of Gibraltar and into the Atlantic.

■ AND WHAT OF THE FUTURE?

There is a tendency to believe that geology has stopped – that somehow the forces that created the momentous changes that we see in the rocks and landscapes around the Mediterranean have come to a halt – but modern satellites can measure them and they are inexorable. With time, the geological forces that destroyed the Tethys Ocean will ultimately claim the Mediterranean Sea too. Morocco will crash against Spain and there will be no need of talk of bridges or tunnels. Tunisia will embed itself on to Sicily, which in turn will ram against mainland Italy. Libya will collide into Crete and stop the southward advance of the Aegean, allowing Turkey to pile into the back of it. Cyprus will ultimately become physically reattached to Turkey, regardless of what politicians decide. In about 34 million years, the leading edge of the African continent will eventually crush against the European continent and the Mediterranean will literally squeeze shut. The mountains that will arise from the depths of the modern sea bed will be far loftier than anything the region has seen in the last 200 million years. The collisions of the past are small fry compared to what will happen when the main bulk of Africa ploughs into Europe. One day our distant descendants may be skiing on the African sea bed.

With the exception of a few scattered ancient cores, most of the rocks that today lie under our feet in the Mediterranean are young – less than 200 million years old, and yet a region whose geology represents less than 5 per cent of geological time has given the western world its earliest civilisations and greatest empires – the Mesopotamian, the Egyptian, the Greek, the Roman. Today, we are more familiar with the history of those civilisations and empires that colonised, conquered and collapsed on its shores than we are with the history of those shores themselves Each culture arose in a different part of the Mediterranean jigsaw and, consequently, on each of their doorsteps was a different set of rocks. Whatever rocks were found beneath their feet would become the building blocks for their monuments and cities.

CHAPTER 3

RAW ROCKS

A LONG AND TURBULENT geological past had left the Mediterranean a complicated place. The moving plates, erupting volcanoes and shrinking seas had left their mark in the bedrock of the region. A wide variety of rocks, almost as wide as the Earth itself has to offer, was scattered across the region. In time, Mediterranean peoples would find more and more uses for the resource they were standing on, but in the early days the rocks provided for their most basic needs. As well as forming the very land that people lived on, the stones beneath their feet were the handiest materials to use as weapons to kill prey (or rivals) or make tools, and it would be rocks that provided our ancestors with their first homes. Such was its importance as a raw material for human development that historians have named our earliest times after it – the Stone Age. The change from 'Old' Stone Age (Palaeolithic), to 'Middle' Stone Age (Mesolithic) and finally to the 'New' Stone Age (Neolithic) is largely the story of humankind's growing sophistication in exploiting rocks, and with the birth of civilisation, rocks would become the foundations of cities and empires.

■ AT THE CUTTING EDGE

From the dawn of the human race, our ancestors learned to give a stone fragment a cutting edge by hitting it with another piece of stone. The earliest stone tools, believed to be more than two million years old, were made in East Africa by flaking volcanic lava. The more uniform the rock material, the sharper the edges of the flake, and the fine-grained texture of lava fragments ensures that they will flake in fairly predictable ways. Around half a million years ago, stone technology really began to gather pace, as humans increasingly became able to control the tool-making process. Palaeolithic people discovered that many hard rocks had natural planes of weakness running through them, called cleavage. Such a rock hit at an angle of roughly 120° to its cleavage planes invariably splits easily. The first geologists were those early craftsmen who carefully determined all the cleavage planes of a raw block of stone and then proceeded blow-by-blow to chip it into a hand axe. Different styles of stone tools emerged, but most had similar functions. Hand axes were used for attacking, cutting and digging, and stone chips for working animal skins. In fact, it was stone that gave our ancestors their cutting edge on the grassy killing plains, providing implements and devices for attacking large prey that allowed the human hunter to concentrate his limited strength to deal localised but lethal blows.

In Europe, the indigenous Neanderthals had long developed stone tools, but 40,000 years ago the first anatomically modern humans, the Cro-

Magnons, began to arrive from Africa via the Middle East, bringing with them more sophisticated stone blade tools. Quite what happened between these two groups of humans still isn't clear, but around 28,000 years ago, in the midst of the last Ice Age (and after around 10,000 years of co-existing), the Neanderthals became extinct. Some scientists argue that the locals were progressively exterminated by the incomers, but most believe that the superior hunting techniques of the modern humans, a function of their better stone culture, probably drove the Neanderthals to ever more marginal environments and ultimately to extinction.

By the time the ice sheets had swelled to their greatest extent, around 18,000 years ago, the ice-free hills and valleys of south-west France had become the most populated part of Europe, and home to an Inuit-type people, the Magdalenians, who hunted large game in the valleys of the limestone hills of the Dordogne. They were so skilled in the making of stone blades that they could get almost a metre of cutting surface from a kilogram of stone, compared with their ancestors' few centimetres. As the Ice Age began to wane, the warming climate coincided with a new revolution in stone tools. Around 10,000 years ago the Magdalenian peoples of southern France and northern Spain now became hunter-gatherers of the Mesolithic as the new varied food sources demanded ever smaller stone blades with specialist functions. Out of this new need came knives and lightweight arrowheads, and tools for engraving, cutting horn or making a notch, as well as harpoons for fishing. Many implements had no direct use, but were simply used to fashion other implements such as the bow, which allowed humans to apply their muscle power from a distance, or the bow-drill, an implement used for boring holes and for making another important tool – fire. As we shall see in later chapters, fire was destined to become a critical feature in the new Mediterranean world.

The demand for specialist stone tools was growing, but not everywhere had the right geological ingredients. Even in earlier Palaeolithic times, areas that did have the right rock – hard and fine-grained with well-developed cleavage – had their hillsides carved into quarries; the local people became craftsmen who exported their products to geologically less-favoured neighbours. Excellent cutting blades came from one widely available source – flint. This is a type of natural glass whose microscopic interlocking crystals of quartz often grow as dense, extremely hard nodules in the chalky limestone strata of southern Europe and northern Africa. However, the most sought-after raw material in antiquity was a much superior and far rarer rock, obsidian. This is a volcanic glass whose texture, caused by the sudden quenching of semi-molten lava, gives excellent flaking qualities and so produces razor-sharp edges.

ROCK SHELTERS

Apart from tools and weapons for survival, early people also needed shelter from the elements and from predators. A temporary shelter could be a skeleton of wood or rushes covered with animal hides, leafy branches and turf, and mud or straw – a basic structure echoed even today with iron and steel frames and prefabricated concrete skins. For a more substantial shelter, though, people would again turn to the rocks around them – which could sometimes provide one ready-made.

The cartoon view of Stone Age Man equates him with Cave Man, but it seems likely that it was mainly the mouths of caves that were exploited. Even then it is unclear whether these subterranean dens were used as permanent dwellings. The fantastic rock-art paintings adorning many of the caverns occupied by the Magdalenian peoples, such as the Lascaux cave in the French Dordogne (which we revisit in the next chapter) appear to have served largely ceremonial or religious functions. Caves may have started out as sacred places, perhaps the first temples, but such natural amenities have long been appropriated for living space. Across the Mediterranean, people have manipulated and enlarged natural rock caverns to form whole villages and towns. Close to Lascaux is the remarkable Rocque de St Christophe, a multi-storey tenement of catacomb-like rooms cut into the sheer wall of a limestone gorge and inhabited from prehistoric times through to the Middle Ages. In Andalucia in Spain, two-storey vaults are carved into craggy limestone rock faces in the Cuevas de Almanzora and are still inhabited; while sixty miles to the west, at Guadix in the Sierra de los Filbares, a crumbling old Moorish town with a vast and extraordinary cave district cut into sandstone cliffs still houses over ten thousand people. Southern Italian hilltowns like Matera also have windowless rock grottoes cut into soft sandstone rocks. These dark and dingy caves now have façades and roofs, but the rock interiors have remained virtually unchanged for thousands of years. Nearly 15,000 people were still living in them in 1952, when the government finally declared them unhealthy and outlawed their use. Today the cave houses are more or less abandoned, the eerie troglodyte suburb left as a curious tourist sideshow to the modern town.

Perhaps the most striking 'living' cave towns are in Cappadocia, central Turkey, a strange and magical moonscape of volcanic rock carved by natural erosion into pinnacles, towers and chimneys. The rock itself is fine ash that was blasted out of nearby volcanoes millions of years ago, but it remains soft enough to dig and develops an armoured crust when exposed to the air. Here people carried on nature's handiwork, sculpting out underground

Opposite: Red sandstone canyons near Roussillon in Provence are made of stratified layers of sand-sized particles laid down as desert dunes and sculpted by erosion into an inspiring landscape.

houses that provided protection equally from summer heat and cold harsh winters on the high Anatolian plateau. Vast underground cities developed, spread over two hundred square miles of volcanic wasteland. The most spectacular is Derinkuyu, whose eight-storey-deep labyrinth of tunnels, rooms, cisterns and ventilation shafts sunk into the soft rock was large enough to shelter ten thousand people. These subterranean high-rises have been occupied off and on for several thousand years, but today they are abandoned. However, elsewhere in Cappadocia people still live in rock-hewn houses, and some villages even offer rock restaurants and rock hotels. With modern comforts of running water, electricity and even television, they are a far cry from the damp and draughty cave mouths of our ancestors.

Whatever use people made of caves in the Stone Age, it seems probable that for as long as they found stones lying around they piled them up to make simple shelters. As early as forty thousand years ago in the Nile valley of Egypt people quarried away at natural rock layers, using gazelle and hartebeest horns as picks to remove cobbles of flint that could be stacked roughly on top of each other. Throughout these Stone Age times, rock continued to be exploited by small groups of people using readily available local materials to produce tools and weapons for their own immediate needs.

What was available depended, as ever, on the geological lottery handed out to them. Almost any rocks that people have on their doorstep will have been used for construction at one time or another. However, of the couple of thousand common types of rock, only a dozen or so are routinely used as building stone. Unsurprisingly, they are the ones that are strong and rigid enough to support free-standing structures and that can be exploited with the technology people have available at the time. However, the properties of these rocks are not uniform, any more than the kind of building they are best suited to: they are characterised by the way in which they have been formed, which in turn dictates the way in which they can be used (a trait we'll explore in the following chapter). Some of the most distinctive monuments of the Mediterranean – the triangular pyramids of Egypt, the rectangular temples of Greece and the circular arches and domes of Italy – owe their shapes in part to the particular properties and limitations of the rocks available.

■ ROCK AND ROLES

There are only three basic categories of rock – sedimentary, igneous and metamorphic.

Sedimentary rocks originate from loose sediment. Many sediments are solid particles that come from broken-down rock and are then slowly compacted

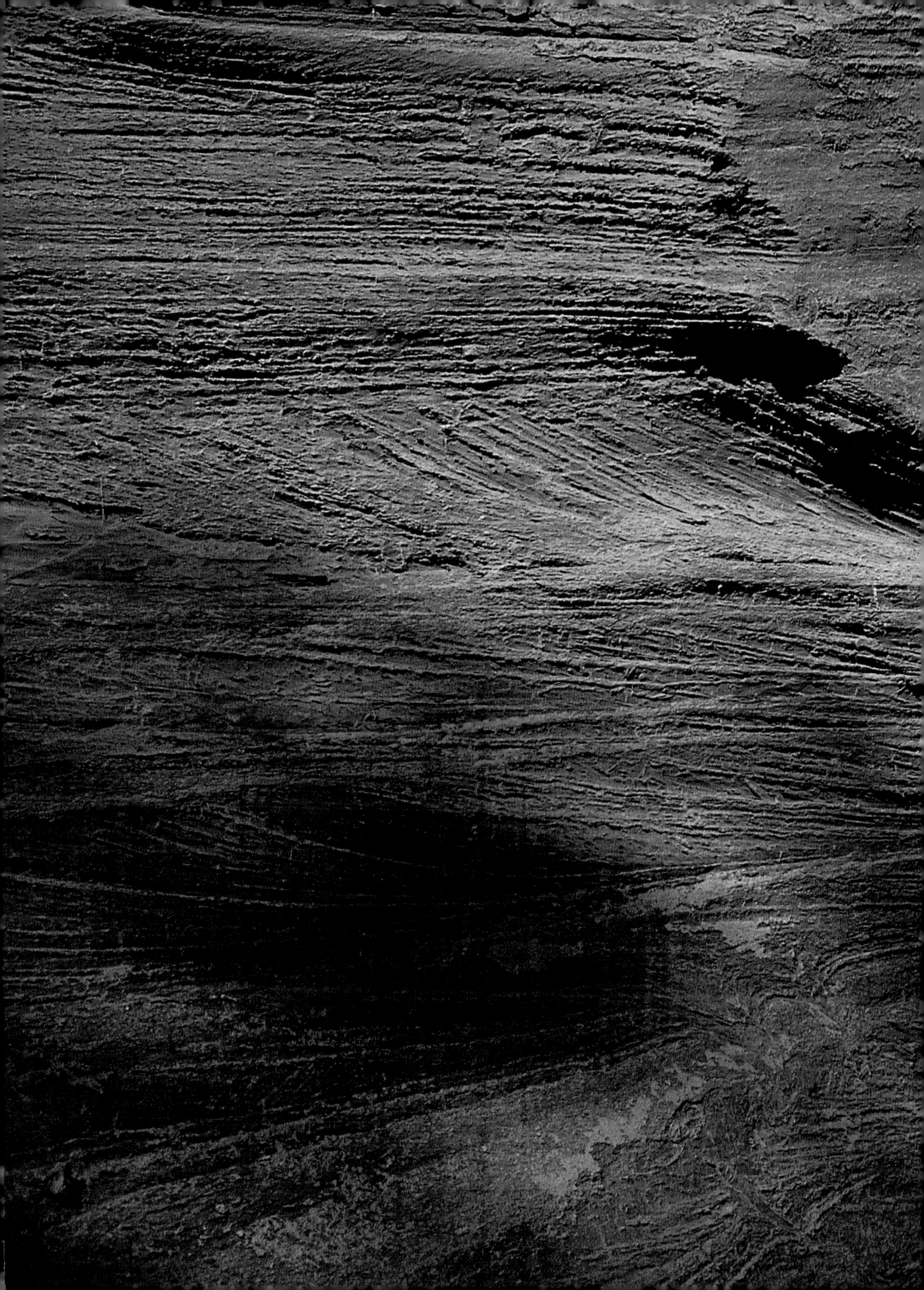

and glued together by thin films of mineral cement dissolved in trapped water. Mud flakes become mudstone (or shale), grains of silt become siltstone, and grains of sand become sandstone. Pebbles or boulders are hardened into rocks called a conglomerate, and if rock fragments are particularly angular they are known as a breccia. Other sediments come from soluble material dissolved from rocks by chemical weathering and later precipitated back into solid form. Many water-dwelling organisms consume dissolved mineral matter to form their shells and hard parts, and when they die that matter is converted into rock. Limestones are the main rock type that forms this way, composed mainly of the mineral calcite (calcium carbonate), but creatures that extract quartz (silica oxide) from water give rise to chert (or flint). The chemical growth of crystals can happen without organisms, such as when salt is left behind when salty water evaporates, forming evaporites like halite (rock salt) or gypsum, or when calcite grows from freshwater springs as tufa and travertine.

Igneous rocks form when magma – a hot molten mush of crystals – solidifies underground or congeals when spewed out of volcanoes. Subtle changes in the chemical make-up of the magma, or in the texture formed as the molten mass cooled, bring about a daunting variety of igneous rock types – some fifteen hundred of them. The familiar granite and basalt are among many others given all manner of wonderfully exotic names, from abessedite and andesite to zobtenite and zutterite. Navigating through this molten complexity needs some handy hints. Colour can be a guide to rock chemistry, since igneous rocks that are light coloured are generally rich in the element silica and are called 'acidic', while darker rocks are more deficient in silica and are called 'basic'. Texture can often reveal how rocks formed, with magmas that have slowly cooled underground forming large crystals, while those that have quenched abruptly at Earth's surface have little or no time to grow crystals, and so appear glassy, like the highly prized obsidian. Simple observations of crystal size and colour are the straws that new geologists cling to as they begin to try to place igneous rocks into the appropriate geological pigeonhole.

Metamorphic rocks are simply sedimentary or igneous rocks in disguise. Heat and pressure have changed their nature, gradually transforming their composition or appearance in a natural makeover. That transformation, or, to use the Greek word, metamorphosis, begins with the changes that the sediment endures as it becomes rock. As the heat and pressure pile on, rocks adapt to the new conditions by rearranging their structure or reorganising their chemistry. Crystals try to line up so that they lie at right angles to the direction of greatest squeezing. Squeeze a mudstone a little and it develops into slate, the clay minerals aligning themselves into a cleavage that allows the rock to split easily. Squeeze and cook slate some more, and it develops a greenish sheen and wrinkles, and it is now a phyllite. Push rocks deeper down into the geological pressure cooker and new minerals grow that are more suited to the higher temperatures and pressures, creating rocks that are now unrecognisable from how they started out. Still deeper and hotter, dark and light minerals bleed out of the rock, forming first a smooth, silvery rock called a schist, then the black and white streaked gneiss (pronounced 'nice'). The most extreme metamorphic rocks are gneisses that are on the verge of melting, and blobs of melt are spread through the rock in what is called a migmatite.

Out of this basic trilogy of rocks comes a huge variety of appearances and textures, and understanding which made the most suitable building material would take our ancestors thousands of years of trial and error. Not surprisingly, the starting point was the most abundant material available, and for the first large-scale building of the first villages and towns, that meant turning to the most common material on the surface of the planet.

MUD, MUD, GLORIOUS MUD

Eighteen thousand years ago Earth was gradually warming up, ending the ice-age conditions that had kept our ancestors in their nomadic lifestyle. By ten thousand years ago, a new warm, wet world was allowing people in the eastern Mediterranean to start sowing their own crops, a revolution that is the focus of Chapter 5. However, the birth of farming meant that people needed permanent shelters and settlements that were close to their fields, not high up in rocky caves or amid stony landscapes. To build these settlements, the first farmers turned to what was sticking to their feet – mud.

Mud is found everywhere – as thin coatings of soil on mountain hillsides, as thick deposits dumped in river valleys, as huge piles building out as deltas into the sea, and as enormous accumulations on the deep sea floor. Flakes of clay broken down from weathered rock are easily flushed through the land and left to settle out as mud in the world's quiet backwaters. Each year only a few centimetres of mud are deposited in lakes, on flood plains or on ocean floors, but over decades, centuries and millennia, more and more mud arrives. The growing weight of the sediment dumped on top slowly squeezes the water out from the flakes of lower layers, forcing the mud particles closer together, and, after thousands of years, this soft mud finds itself buried several hundred metres and tightly compacted. Several metres of loose mud will have become compressed into a metre of rock.

The numerous delta plains and broad river valleys around the Mediterranean have long ensured that there was plenty of mud about. In Mesopotamia – 'the land between the rivers' of the great Euphrates and Tigris waterways – new settlements were emerging and expanding across the flat river plains, where people were able to harness water to irrigate fields and till the soft muddy soil. Here, mud was everywhere. Quickly people discovered that it was easy to work with and wonderfully versatile. Mud dug straight from the ground could be shaped by hand, squeezed into rigid moulds to form blocks, rammed between wood and rushes to form walls, or plastered on the outside of a structure to form a smooth, waterproof shell (so-called 'wattle and daub', where mud is the glue that sticks together a mixed set of other materials). Then, just as heat and pressure of the Earth transformed mud into mudstone, so the baking Mediterranean sun ensured that soft river mud could quickly be hardened.

Mud is such a useful building material that it is still used today. In Morocco, the Middle East and even Spain, sandy, muddy soils are mixed with water to make great earth bricks. In Iran, the use of this 'adobe' construction is widespread, with mud bricks being clad, or rendered, with mud, straw and even animal dung to provide a smooth surface that reduces wind erosion. In northern Syria these adobe bricks are stacked in an offset fashion to produce an egg-shaped, or corbelled, roof that re-radiates heat.

One of the earliest mud towns that we know about was Catal Hoyuk, which flourished nine thousand years ago in the plains of central Turkey. Houses were made with wood frames filled in with mud bricks, reeds and plaster; bundles of reeds capped with a mud coating formed the flat roofs. Thousands of these mud-brick houses were crammed together in a town without streets; hatches in the ceilings allowed people to move around on the roofs. The mud brick meant that individual buildings didn't last too long, but moving house was easy. When a house had started to crumble or outlived its purpose, the inhabitants simply demolished it (usually with their dead relatives entombed inside), and then built afresh on top of the ruins. With something like eighteen different levels spanning a period of twelve hundred years, Catal Hoyuk was definitely an upwardly mobile community.

The first cities emerged in flat plains devoid of forests and rocky hills, so without wood or stone they were founded on mud, although this is hardly credible from what is left today. In Mesopotamia, little trace is left of the cities built by the Sumerians seven thousand years ago in the southern marshlands, or of those established by the Babylonians five thousand years later on the central flood plains. They crumbled away. It was a different story in the mountainous north where the third great Mesopotamian empire,

Assyria, had stone available for building and so has left behind great stone fortifications, albeit clad in mud brick. Ironically it was not stone but mud that gave the Mesopotamians their most distinctive form – the ziggurat. This stepped pyramid made of rubble and mud brick was probably a form born out of necessity. The short life of mud brick meant that constant rebuilding was required, so each successive rebuilding took place on the debris of previous temples. Undoubtedly the mound also had a symbolic significance – artificial mountains in a flat world – and far to the south, another civilisation would take this form to new heights.

The Egyptians, like the Mesopotamians, relied on mud. It was washed down from the highlands of Ethiopia and East Africa along the Blue and White courses of the Nile, which merged in lower Egypt into a single raging river. For three months of the year the great river would break its banks and fresh organic-rich black mud would spread across the flat adjacent plain. As the waters retreated, the fresh mud would be slowly baked hard in the sun, turning it from dark brown to grey. Blocks, made from it, strengthened by reeds, straw and dung, would remain the preferred building material until the time of the first pharaohs, about 3000 BC, when they discovered brick.

The hardening of mud through heating in kilns to form mud brick was an ancient observation shared with the Mesopotamians, but a lack of trees in the Nile valley meant wood-fired kilns were an expensive business. It was much easier simply to leave it to the baking sun. With plenty of sunshine and with the raw material being annually replenished by the Nile floods, mud-brick architecture flourished in ancient Egypt, where adobe buildings ranged from humble starter homes to huge palace complexes, many rising to considerable heights.

One of the most remarkable of Egypt's mud cities was Amarna. Situated in an open desert plain close to the Nile nearly three hundred kilometres north of Thebes (modern Luxor), Amarna is a spectacular example of the transience of mud-brick architecture – a city founded, built, occupied, abandoned and destroyed in a single generation. It came about around 1350 BC, when the king, Akhenaten, decided to abandon Thebes, with its strong connections to the god Amun, and to found a new capital city at the heart of his empire under a new deity, the divine sun-disc, the Aten. With the new city needing to be built quickly, mud brick, which could be produced readily and rapidly, provided the obvious answer. Although the most important public buildings, the temples, were made of stone, houses, villas and palaces were quickly constructed out of the local mud; the Great Palace alone had an enormous hall crowded with over five hundred brick columns. In the times following the death of Akhenaten twelve years later, the city was abandoned as a later king, Tutankhamun, despite being born in Amarna, restored the traditional state religion and relocated his capital back to Thebes. The temple stonework was pillaged, leaving the monumental masses of brick-work to gradually crumble into the desert – a sad testament to a failed early experiment in town planning.

The first major monuments in ancient Egypt were mud-brick tombs, known as mastabas. These structures began as simple mounds of earth raised over graves dug for royalty and nobility. Similar in style to the Mesopotamian ziggurat, like them they often didn't last too long. For the ancient Egyptians, their beliefs in death, fertility and regeneration meant that they needed tombs for their pharaohs that would last not just for a few centuries, but for eternity. Their mud-brick tombs were anything but immortal. What the Egyptians needed was stone – and they got it from the most unlikely of places: tiny spineless sea creatures.

SKELETON STRUCTURES

Just like the Mediterranean Sea today, the Tethys Ocean was teeming with marine creatures. There were all sorts of fish and reptiles, but it was the other animals that made up the abundance of marine life. The geologically important animals were the ones without backbones – the invertebrates – like crabs, clams and starfish, but also the billions of tonnes of microscopic plankton. Although spineless, all these animals had hard body parts, generally made of calcium carbonate, and when these creatures died, any soft tissue would have decayed away, or been eaten. What eventually came to rest on the ocean floor were layers and layers of these inedible calcium carbonate body parts, which accumulated over millions of years. Slowly but surely, this muddy graveyard was squashed and compacted by the accumulating weight of all the new dead animals that were continually being deposited on top. This pressure, along with time, meant that the sea bed of the ancient Tethys Ocean was a giant rock factory, transforming what had been living, breathing sea creatures into limestone.

Around sixty million years ago, the northern and southern fringes of Tethys were covered by warm shallow seas, but by the time of the ancient Egyptians, that limey sea bed had been raised high and dry. So, when the pharaohs wanted to build with stone, it was limestone that they found on their doorstep. They switched from mud brick to stone during the early part of the so-called Old Kingdom (2650–2150 BC), five centuries of more or less uninterrupted economic prosperity and political stability. This boom period gave the pharaohs the time and money to immortalise themselves in true monumental grandeur, rather than in the mud-brick tombs of their ancestors that were already showing signs of natural wear and tear. During what would be ancient Egypt's 'golden age', the royal heart of the kingdom was centred on Memphis in the north. Here, and extending as far south as ancient Thebes, limestone strata occur almost continuously in the Nile valley and in the desert plateau to the east and west. Over eighty ancient limestone quarries litter this stretch of the great river, most nestling close to the temples and pyramids that they were used to build. It was perfect for the unprecedented demand that was to follow.

The switch to limestone was gradual. For a few decades, pharaohs planning for the afterlife had begun to incorporate stone in their mud-brick funerary complexes, but as limestone blocks began to be used more and more, early Egyptian builders seized on the chance to get beyond the limitations of the traditional rectangular tomb. They began thinking beyond the box.

PYRAMID SCHEMES

Around 2650 BC, Imhotep, the architect designing a tomb at Saqqara for the pharaoh Djoser, took what started out as a standard rectangular tomb, and fashioned it into one that had six steps, so that it looked like a stepped pyramid. Not only that, but the building was constructed entirely of limestone blocks. This 'step pyramid' is not only the earliest pyramid in Egypt, it is the world's first successfully completed stone building. Its designer – the inventor of building in stone – got his reward in heaven: Imhotep was made a god and given his own stone tomb.

Stone building really took off during the later reigns of the Old Kingdom pharaohs Snorfu and Menkaura, when around nine million tons of limestone were quarried for funerary complexes, first at Dahshur and then at Giza. Initially, Imhotep's stepped pyramid form became the de rigueur funerary accessory for any self-respecting immortal ruler, but gradually this evolved into the smooth triangular shape associated with Egypt's pyramids today. The first attempt wasn't particularly encouraging. At Maidum, a stepped stone pyramid was converted into a 'regular' pyramid but, having been built on desert sands rather than on solid bedrock, it gradually collapsed. Later, in the reign of Snorfu, builders constructed a sloping pyramid at Dahshur. Although the lower two-thirds was built at an angle of 52 degrees, the builders, perhaps worried by the unstable form of the previous one at Maidum, reduced the upper slope to a more stable angle of 43 degrees, creating what is now known as the 'Bent Pyramid'. Rather than its intended height of 130 metres,

Tourists caught up in sandstorm at Giza, overlooked by the famous Sphinx and the only one of the Seven Wonders of the Ancient World left standing – the Great Pyramid of Giza.

about the height of St Peter's Cathedral in Rome, the redesign reached to only 110 metres, still equivalent to St Paul's Cathedral in London and twenty metres taller than the Statue of Liberty. Their next attempt at Dahshur, the Red Pyramid, was built completely at an angle of 43 degrees, but to get it to rise to a decent height (just over a hundred metres tall) it was given an enormous base, around 220 square metres. Throughout Snorfu's reign the best pyramid shape and dimensions were road-tested, and by its end all the engineering elements were in place to construct the biggest pyramids of them all. Combining the base size of the Red Pyramid with the initial slope angle of the Bent, they achieved two stone pyramids that were over 140 metres tall. They were destined to be the only wonder of the ancient world that fate would leave standing – the Great Pyramids of Giza.

Opposite: Constructed around 2650 BC, the 'Step Pyramid' at Saqqara is not only the earliest pyramid in Egypt, it is the world's first successfully completed stone monument. What began as a simple square tomb was transformed gradually into six steps to set the shape of Egyptian monumental architecture for the next few thousand years. Perhaps the first example of thinking beyond the box.

The pyramids of Giza are built from limestone, though not from the limestone bedrock immediately on their doorstep. The limestone underfoot at Giza is too soft for serious building purposes. It is so soft, in fact, that you can easily cut straight into it, and that is exactly what ancient stone masons did when they carved a sculpture seventy metres long and twenty metres tall straight out of the rock. Today, rather battered and bruised and with a dubious nose-job, that rock carving is one of Egypt's most famous landmarks – the Great Sphinx.

The Giza pyramids themselves were constructed from two different limestones. The outer casing was a pale grey, fine-grained limestone that made it both attractive and durable. However, for the enormous blocks that built the interior, the Egyptians turned to an older, harder limestone that formed a cliff, several hundred metres high, along the edge of the Nile valley. Transporting blocks of limestone from so far away was a great way of soaking up spare labour. For those three months of the year when the farmland was covered by Nile floodwaters, thousands of agricultural workers had nothing to do, except potentially sit around and get restless. Instead, what better way to keep them busy than to build a very big structure out of stone? It's reckoned that 100,000 men worked in three-month shifts, labouring for twenty years to build the Great Pyramid alone. Pyramids were a lifetime's preoccupation.

Just why did the Egyptians choose to build those earliest stone monuments as giant triangles? For some, the notion that Egypt emerged as a mound from a primordial sea is reason enough to see pyramids as their man-made mounds. Certainly in the relatively flat landscape of the Nile plain and the adjacent desert plateau, these stone triangles stood out. In fact, in the shape of their early pyramids, the Egyptian triangles were mimicking that ultimate stable shape – the mountain. Both mountains and pyramids have to

The Great Pyramid of Khafre (Cephren) at Giza near Cairo was constructed out of the remains of ancient sea creatures. Limestones that formed on the floor of the Tethys ocean provide the enormous blocks for the great monument as well as the decorative stone cladding for its polished outer casing, most of which has now been removed.

act against gravity. To hold up their own weight, a pyramid form is the most efficient way of spreading the load. The few stones at the top only have to support their own weight, but lower down where the weight increases there are more blocks to share the load. In fact the height of most Egyptian pyramids is two-thirds the length – which turns out to be the most efficient mathematical shape for distributing weight. So, if builders are going to stack blocks, then the simplest and most stable structure is just a set of walls angled towards each other to provide mutual support. In other words: a pyramid.

GIANT SANDCASTLES

The trouble with pyramids was that they were a security nightmare. They virtually shouted out 'This place is stuffed full of treasure', so the triangular tombs were routinely raided and robbed. The ancient Egyptians needed a better way to get their kings to heaven without being desecrated on the way. In the 18th Dynasty, around 1500 BC, when the ancient capital moved from

Stratified layers of Nubian sandstone make an ideal rock in which the Egyptians could carve their fantastic hieroglyphics.

Memphis in the north to Thebes in the south, geology provided them with a new way to venerate their dead. In the south they encountered a different sedimentary rock. Between Memphis and Thebes was limestone territory, but the land of Upper Egypt south of Thebes was sandstone country. The ancient Egyptians decided to use the limestone hills north and west of the new capital as a vast natural city of the dead, a necropolis of rock-cut tombs burrowed into the cliff walls – the Valley of the Kings. Then, below on the

Opposite: Horizontal sedimentary layers feel the stress of distant collisions by breaking across vertical cracks or joints, such as those that here cut sandstone strata in central Arabia. Such regular metre-spaced joints ensure that flat-lying limestones and sandstones are often available as pre-cut ready made building blocks. [Dave - Perhaps this should have a wider view so that the figure can be seen for scale]

plains near the Nile, they constructed temples to commemorate the pharaohs resting in the rocky hideaways. These were built out of the local sandstone.

Sixty million years ago, the rocks beneath what is today Upper Egypt were soft sediments building up close to the southern shore of the Tethys. Rivers draining the mountainous African interior built up kilometre-thick piles of sand and silt, exactly as the Nile delta does today at the Mediterranean Sea. The ancient sandstone was found almost continuously in the Nile valley from Thebes south to northern Sudan, as well as on the nearby desert plateau. Over thirty ancient quarries are found along this stretch of the great river, most excavating a geological formation called the Nubian Sandstone. The people of southern Egypt had been building tombs out of this sandstone from early pharaonic times, but the first great monuments began to appear during the 11th Dynasty, 2000 BC, when deceased pharaohs and their deities began to be worshipped not in pyramids but in temples. As kings came and went, temple complexes were built, enlarged, torn down, added to and restored, leaving an apparent chaos of walls, obelisks, columns, statues and decorated blocks. An explosion in sandstone construction came with the move of the royal administrative centre to Thebes. The Egyptian builders of the 18th Dynasty found sandstone to be superior to limestone in terms of the size and strength of the blocks that could be extracted from quarries; furthermore, it was more easily worked and better for shaping and for carving hieroglyphics. Sandstone was quickly adopted as the building stone of choice for the entire kingdom, north and south. Limestone, more durable and capable of being intricately carved, was relegated to sculpture.

Sandstone reached a pinnacle of glory in Karnak, near Thebes. Constructed and deconstructed over a period of about 1,500 years, during the height of the Theban royal court, it was the largest and most important temple complex in the whole of Egypt. The outer walls and precincts are made of mud brick, but its central enclosure is a feast of sandstone architecture, particularly the magnificent Amun Re Temple and its awesome Great Hypostyle Hall. Built by Pharaoh Seti I in the 19th Dynasty (around 1280 BC), the enormous hall has over 130 towering sandstone columns holding up a brightly painted stone roof; it is big enough to house the cathedrals of both St Peter in Rome and St Paul in London.

Both limestone and sandstone were used, over several thousand years, for nearly all the major monumental structures in ancient Egypt because, again, they were the most abundant and easily accessible rock on that civilisation's doorstep – but it was the properties of these sedimentary rocks that were key to the building styles that developed.

The Great Hypostyle Hall in the spectacular Temple of Karnak, near Luxor in Egypt. The dense hall of columns is supposed to represent a primeval swamp of papyrus, but it is also a remarkable celebration of the use of sandstone as a building material.

CRACKING IT

Essentially, sedimentary rocks can handle being squeezed together (compression) – not surprisingly: piled up on the sea bed, sandstones and limestones have already been squashed for millions of years. Even the bottom block of a large pyramid or column at Karnak is not going to be under as much weight as it was hundreds of metres below the bed of the Tethys ocean or the Nile delta. However, while sedimentary rocks take readily to compression, they can't stand tension (being pulled apart). If a layer of sandstone or limestone is forced to bend, the material on the inside of the bend undergoes compression while the outer edge is stretched. It takes only a small amount of tension on that outer edge to prise open the 'glue' between the rock grains. When that happens, cracks appear – just as, on a rather smaller scale, a chocolate bar will show a set of breaks forming on the stretched surface when it is bent. Even though the compressional side of the layer is perfectly comfortable with the stresses imposed on it, the layer as a whole snaps, creating a tensional crack, or joint.

Like a furrowed brow, joints are the rocks' first outward signs of strain. They are the simplest and most common geological structure and occur in virtually all rocks, but particularly in well-layered sedimentary rocks. Stresses from distant land movements and collisions can ripple for hundreds or even thousands of kilometres through such rigid strata, forcing the rock layers to flex, and consequently crack. In the layers the Egyptians used for building stones, the cracks are roughly every two to three metres. Just the right size, it would seem, for the stonemasons to extract ready-made building blocks.

The trouble was that what happens in nature also happens in buildings. The tendency for sedimentary layers to bend and crack under tension put severe limitations on what the Egyptians could build. For a start, roofs were a problem. To make the pillars of the Great Hypostyle Hall, the Egyptians cut the sandstone into blocks, which they rounded off into discs. Then, by stacking the discs on top of each other, they created huge tall columns. Across the top of the columns they placed great sandstone roof beams. At either end of the stone beams the weight of the rock was held up by the columns, but between them the slab sagged under the weight of gravity, forcing it to bend and the roof slab to crack. To get round this, the Egyptian builders were forced virtually to fill the hall with the giant pillars to support the heavy roof slabs that would have once covered the whole ceiling. This dense hall of columns of the Great Hypostyle Hall at Karnak is supposed to represent a forest, or primeval swamp of papyrus, but it is difficult to see how else the builders could have constructed it with the limitations of the sandstone blocks to hand.

By pushing the properties of sedimentary rocks to the limits, the Egyptians had established a new fundamental way of holding up buildings. They were the first to get away from simply stacking stones on top of each other and, in doing so, they had invented one of the classic architectural forms, the post-and-lintel. This combination of a pair of vertical columns capped by a horizontal cross beam allowed them to start building stone roofs. Four and a half thousand years later we can still admire their achievements – and these achievements were not lost on their neighbours in the ancient Mediterranean world either. Foreign mercenaries and traders, inspired by this awesome monumental stone architecture, would gradually take stone-working skills home, and one particular civilisation, the ancient Greeks, would use lessons learnt from the Egyptian stone masons to inspire them to go one better.

■ FINDING THEIR MARBLES

To the traveller approaching the Greek capital of Athens by boat from the south, the Temple of Poseidon at Cape Sounion is a pretty spectacular 'Welcome' sign. Set on a burnished rocky headland jutting into the Aegean sea, its gleaming white marble columns dazzle against the blinding blue of sea and sky. The effect was intended to be both immediate and enduring, for in contrast to the introverted Egyptian monuments, ancient Greek temples wore their beauty and splendour on the outside, bearing friezes painted in garish reds, greens and golds and being adorned with statues inlaid with coloured gemstones. Deliberately placed to be in harmony with their landscape, these were monuments for the gods above, not for dead kings within.

To achieve this, the ancient Greeks concentrated on perfecting the rectangular post-and-lintel temple form they had borrowed from their Egyptian neighbours – but what allowed them to take this to new heights of sophistication and architectural splendour was a rock that the Egyptians just didn't have under their feet: marble.

At first, it hadn't occurred to the ancient Greeks to build with marble. Initially, they hadn't needed to. The geological lottery isn't just about what rocks are available. The Greek mountains created by the tectonic collisions fifty million years ago efficiently captured rain swept in off the Mediterranean, making Greece far wetter than the land of the pharaohs. Forests covered the Greek mountainsides and this timber was used for building: constructing houses and temples with wooden columns and roofs. Early

A marble column at ancient Ephesus in western Turkey. Marble was the preferred building material for the ancient Greeks because heat and pressure had fused the limestones beneath their feet into a stronger rock in which original weaknesses like cracks were erased, making it ideal for building and sculpture.

builders also used the wood to fuel kilns in which they would turn their deposits of mud into stronger, fired bricks. For the first time, terracotta roof tiles appeared; these tiles were much more waterproof and allowed lower roof angles than reeds and thatch, but there was one big problem – they were much heavier.

Wood, although not as good as stone in compression, is as strong in compression as it is in tension. In other words, wooden roof beams, although not highly efficient at holding weight, could stretch across wide gaps between walls. However, if the ancient Greeks built their temples beyond a certain size, the weight of the terracotta roof tiles would be too much for the wooden pillars and brick walls to support. Their buildings seemed to have reached a limit just when their civilisation was on the up. The difficulties with terracotta roofs meant that wood persisted as a roofing material even in the later stone temples, but its susceptibility to fire is probably the main reason why many have been left to us as roofless ruins.

From around 800 BC, the wooden pillars and beams of Greek temples began to be replaced by stone columns and blocks. Sedimentary rocks like sandstone and especially limestone were all around in the Greek landscape, but unfortunately the millions of years of mountain building had left them with contorted and broken strata. Greece was rather like a geological motorway pile-up – the flat layers of pre-cut blocks that the Egyptians had on their doorstep were hard to find. However, in the heart of the ancient mountain belt the sediments had been pushed down to depths of forty or fifty kilometres and heated to temperatures of over 400°C. In this deep pressure cooker, the sedimentary rocks turned into metamorphic ones.

Limestones, which are almost entirely composed of calcium carbonate remains, don't have too many options when it comes to being metamorphosed. When it is cooked and squeezed, calcium carbonate regrows into a tightly interlocking mass of calcite crystals, giving a new rock – marble. Just as something plastic or metal softens when it is heated or compressed, so marble deep underground can flow, and cracks that it developed when it was cold limestone become fused together. If the original rock is made only of calcium carbonate, then it will turn into a pure white marble, but generally impurities infuse the marble with the swirls of colour that make for beautiful decorative stone. Because the new calcite crystals are roughly the same size and because original weaknesses in the rock are erased, the marble can easily be cut and polished, making it a favoured stone for ornamental building or sculpture.

Marble is a pretty tough rock to mine – expensive in terms of both time and labour. Not surprisingly, it wasn't used extensively for building until the fifth century BC when Athens had its real boom time. The city was already prosperous on the back of its famed pottery and its silver wealth, but a war chest collected from the Greek city-states to fight the Persians was siphoned off by the Athenian leader Pericles to finance a colossal building programme. Suddenly, and much to the annoyance of its neighbours, no expense was spared for temples and buildings destroyed by the Persian invaders. The foundations for the new buildings were mostly of limestone, but the stone used for the construction of the Parthenon and many of the great buildings of this time was marble. The purest of the marble was from Mount Penteli, a 20-kilometre-long lump of metamorphic rock north-east of Athens.

As well as in the hills around Athens and eastern Greece, marbles are found in an arc of metamorphic rock that swings through the Cycladic islands and into western Turkey. Islands like Paros were renowned in antiquity for snowy-white marbles that rivalled those of Penteli, but the largest swathes of marble were in Turkey. As ancient Greek colonies settled on this opposite side of the Aegean sea, their new cities quickly grew rich on trade. The new inhabitants of this coast – ancient Ionia – developed special trading links with the Nile delta, and their envious eyes turned towards the enormous Egyptian temples. Up until this point the Greek temples had remained relatively modest – if perfectly formed – but, blessed with an enormous backyard of quality marble, people began to build big.

The ancient Greek stone masons of Ionia found that the fused, reconstituted nature of marble gave them longer blocks that were strong enough to stretch across gaps up to 50 per cent longer than the Egyptians had been able to achieve. Essentially, the superior strength of the material allowed their doorways to get wider and their columns to get thinner. In came the 'Ionic' column – slimmer, lighter, more delicately sculpted, and, crucially, taller than the chunky 'Doric' columns preferred by the mainland Greeks. Later the chunky Doric and slimline Ionic columns would be joined by an even more ornate style championed by the Romans, the 'Corinthian'.

Ionian temples soon became larger and more spacious than those of the Greek mainland and gave the ancient world two of its Seven Wonders. One was the huge mausoleum of Halicarnassus (modern Bodrum), built in 355 BC of mainly limestone blocks, but sheathed in marble imported from Athens, the Cyclades and central Turkey. Arguably the greatest of all the ancient wonders, however, was the Temple of Artemis in Ephesus, the largest temple in the Greek world and the first monumental building to be constructed entirely of marble. Having learnt the art of working in stone from the Egyptians, the ancient Greeks had discovered that marble enabled them to build the most beautiful rectangles in the world.

Ionic columns, such as these at the ancient Greek city of Priene in western Turkey, were generally slimmer, lighter and taller than their chunky counterparts on mainland Greece, a testament to the new improved use of local marble.

■ HOT ROCKS

Captivated by the beautiful marble colonnaded temples of the Greeks, the Romans would make marble *the* building stone of the Mediterranean. Successive emperors brought the stone back to Rome as loot or through trade, beginning with Augustus in the first century BC – it is he who is said to have found Rome a city of brick and left it a city of marble. The Eternal City grew rich on the revenues of the marble trade and, as the empire spread into North Africa and the Middle East, marble followed. The movement of stone was comparatively cheap and easy by sea (though not by land), and under the Romans marble became sufficiently prized to be widely traded across the Mediterranean. By the second century AD, marble hewn from mountains in Turkey was being lavished on new coastal cities like Leptis Magna in Libya. Perhaps for the first time in the Mediterranean, people

Looking into the jaws of the south-east summit crater at Mount Etna, Sicily. How violently material erupts out of the crater depends on how quickly the material has risen to the surface and how much gas is contained within the approaching semi-molten magma.

weren't building with what was on their geological doorstep.

However, the marble splendour of the Roman world is, literally, a façade. At face value, many of Rome's temples and colonnaded buildings could just as easily be in Athens – but the marble is only skin deep. The truth is that the Romans were a practical people. With a rapidly expanding empire, and a swelling population at home in Rome, they needed to build faster and bigger than any civilisations that had gone before. They needed mass production of building stones, and what gave them the edge were volcanoes.

As well as Mount Vesuvius and the volcanic craters that ring the Bay of Naples, active volcanoes occur to the south on the Aeolian Islands, and there is Mount Etna on Sicily. To the north, Rome itself is surrounded by seven volcanic hills, now quiet but part of a once violent explosive landscape that stretched up the eastern edge of Italy as far as Tuscany. Today many of those fiery centres may have gone cold, but the rocks that they spewed out litter vast areas of what would become the heartland of Imperial Rome, and it would be these igneous rocks that would give Rome the edge. However, the type of igneous rock that you have on your doorstep depends on the journey that it has had to the surface.

The source of Italy's hot rocks lies tens of kilometres deep. As the oceanic floor of the African plate is pushed down into the mantle beneath the Italy, the cold ocean crust is heated up and squeezed. Water is released from the ocean rocks and the wet sediments that cling to them. The presence of water at such depths has a dramatic effect – it lowers the melting point of the surrounding semi-molten mantle rocks and allows them to melt. Now more buoyant than normal mantle rock, the melted blobs of rock – magma – begin to rise up through cracks and faults in the crust above.

Initially, the rising blobs of magma are identical to the mantle material from which they emerged, and are referred to as basaltic, but as they rise through the crust, they begin to take from the continental rocks more and more of the lighter elements, especially silica. As the magma gets rich in silica, it becomes increasingly thick and syrupy, and changes into a granitic melt. Granitic magmas don't travel well, finding it difficult to ooze through the narrow pathways in the crust. As they struggle upwards, the magma cools, eventually 'freezing' into solid rock – granite. Trapped deep down in the earth, granites must wait for millions of years for the slow hand of erosion to gradually strip off the rocks above them so that finally they can see the light of day.

Granite's bright, chunky appearance has long made it a popular ornamental building stone. The crystals that grew large by slow cooling in the bowels of the Earth are easy to see with the naked eye, and the colourful mineral ingredients and coarse texture make granite an obvious ornamental rock. In the ancient world, everyone from the Egyptians to the Chinese used it for columns or for cladding their buildings, but there is a problem – it is extremely difficult to get out of the ground. Granites lack the natural layers and fractures that sedimentary rocks have, and their structure of large hard crystals that have slowly grown to interlock with each other make them a stone mason's nightmare. Because of these drawbacks, they've never really made it as a major building stone. Generally, unless it's for something really special, granites are just too much like hard work.

Basaltic magmas may also 'freeze' if the journey to the surface is too convoluted, but if not, their more fluid nature means they are able to move to the surface faster. Their speedy rise is helped by gases like carbon dioxide, water vapour and hydrogen sulphide (the one that smells like rotten eggs), which at depth are kept dissolved by the high pressures. Close to the surface

Igneous rocks exposed in St Peter's Square in Rome. The pink-coloured bollard is made of granite, whose coarse interlocking crystals originally grew as the semi-molten magma slowly cooled deep within the planet's crust. The dark grey paving stones are made of basalt, which contains small crystals that cooled within a lava flow at the planet's surface before they had time to grow.

A basaltic lava flow formed during the 1991-1993 eruption of Mount Etna which buried fields and houses and came close to the town of Zafferana on the volcano's upper flanks. Despite the strength and durability of basalt, the fractured, rubble-like texture of fresh lava flows make them a difficult material to exploit for building.

the pressure gets less, and these gases start to leak out. What happens in the final few kilometres of rise is crucial for the type of rock formed. If the magma rises slowly, then the bubbles have time to escape quietly and by the time it reaches the surface the magma oozes out as a relatively harmless lava flow, gradually cooling into basalt rock. However, if the magma rises quickly, the gas will be trapped and when it reaches the surface the lava will explode out in a large eruption cloud. As the cloud billows, the coarser material falls first, from boulders to pebble-sized lumps. Later, the lighter material rains down – frothy volcanic foam that solidifies as pumice, and fine dust called ash.

People have been building with what has been thrown out of volcanoes ever since they first started living alongside them. Basalt has been widely used, the hardened eruption streams serving as natural quarries. The dull black rock is relatively easy to extract, being broken up into blocks by large fractures. Most of these arise from the cooling down of the flows, which forces the rock to split into roughly hexagonal-shaped slabs, much like the cracks on the crust of a drying bed of mud. The basalt is tough and heavy, and its rough interlocking slabs could be detached and then reassembled jig-saw-style as early paved roads, such as those that can be seen flooring Trajan's Market and the main Forum in Rome. Further south, in Naples, many of the modern streets, and some of the buildings, are built out of lava blocks hewn out of the ancient flows that swept down from nearby Mount Vesuvius. For all its durability and ease of removal, however, basalt was not destined to become the stone that built the Roman world – the fractured lava flows were just too irregular and too localised to furnish an empire.

So, neither of the two main types of igneous rock – basalt and granite – is much good as a building stone. In fact, the material that would change the buildings of the ancient world would come not from hard igneous rock at all. Instead, it would come from the softer by-products of volcanic eruptions – ash. At first glance, volcanic ash doesn't seem to be up to much as a building material. It is soft enough to crumble with bare hands and so light it floats on water – but over time the soft ash is compacted and cemented into a harder, consolidated material which geologists call 'tuff' (and which ancient Roman builders referred to as 'tufa'). For a million years the Italian volcanoes had been throwing out vast quantities of this ash, making it the most abundant and easily quarried material on the Imperial doorstep.

The Romans discovered the usefulness of tuff when they built the harbour and port of Puteoli (modern Pozzuoli) on the Bay of Naples. At first, they used the tuff along with lava and pumice to crush and pack in behind brick walls, but they soon discovered that when the tuff was ground down into a powder, the so-called 'dust of Puteoli' or pozzolana, had strange but wonderful properties. Added to dry lime, the silica and aluminium in the powdered volcanic ash reacted to form a powdered cement.

Cement itself was nothing new. In fact it has been around for almost as long as people have been building permanent structures. Nine thousand years ago early Neolithic peoples made a crude lime cement by roasting limestone cobbles in a kiln for several days to form a lime-rich residue (at temperatures of 900°C carbon dioxide burns off from the calcium carbonate of the limestone to leave calcium oxide, or lime). The lime powder would then be mixed with water to form a paste which, when left to dry in the

open air, takes in carbon dioxide from the atmosphere to produce a synthetic calcium carbonate rock – lime cement. For building, an aggregate material can be added to the cement as a binding agent. The addition of fine sand to wet lime cement forms plaster, of coarser rock particles forms mortar, and of lumps of rock makes concrete.

A basaltic lava flow formed during the 1991-1993 eruption of Mount Etna which buried fields and houses and came close to the town of Zafferana on the volcano's upper flanks. Despite the strength and durability of basalt, the fractured, rubble-like texture of fresh lava flows make them a difficult material to exploit for building.

Lime cement has long been used in the Mediterranean. From earliest times mud walls were protected from the elements by a moist coating of the lime plaster, which was found to harden the walls to the effects of wind and rain. Whitewashing of outer walls with the reflective lime paste was also good for cooling interiors, a practice that seems almost obsessional in the Greek Cyclades. The ancient Egyptians and Greeks used lime cement in a very different way, in waterproofing baths, cisterns and floorings. However, unlike the cements that the Egyptians and the Greeks had used, Roman pozzolanic cement didn't need water to be made, and so did not need to dry out in order to harden. Instead, it was the addition of water that made the cement harden.

The fact that pozzolanic cement set in water opened the way for the Romans to build underwater concrete structures. Around the Empire, artificial stone harbours could now be built quickly along coasts that lacked natural rocky promontories. At Caesaria, Herod's capital in the Levant, a particularly straight and soft, sandy coastline was unsuitable for a natural harbour, so Herod imported cement from Pozzuoli, two thousand kilometres away, to build an enormous concrete harbour, far larger than would be possible with quarried stone. Of course, importing pozzolanic cement was an expensive business, and sometimes there were just as good local equivalents. The Romans also used ash found in the Aegean volcanic islands such as Santorini and Melos, both areas that still produce the same material today. Although natural pozzolanas are derived from volcanic products, some clays are also rich in silica and aluminium, so where volcanic ash was not available, crushed terracotta could also be used to give a pozzolana effect.

There was so much of this ash lying around that the Romans began to use mortar as bricks in the interior walls. The volcanic tuff was also a far better aggregate for making concrete. The tiny holes in the hardened ash that made the tuff so light and soft provided the key to this secret strength. The wet lime could get inside the porous structure and once there it set hard, gluing the two substances together and locking the rubble and rocks within the concrete gel. This means that large blocks of cement may be pre-formed or cast in situ without the concern that air needs to reach the centre of the block to permit hardening. By pouring this concrete into wooden trays or moulds before it set, the Romans were able to mass produce the first very

strong, man-made pre-cast blocks.

The Romans no longer had to rely on nature to supply their large building stones; they could – using a natural product – make their own. What's more, they were lightweight and water-resistant. The Romans used their concrete stones and bricks to make the frames of buildings, but then clad the surfaces with the marble they so admired in Greece. Instead of using the large stone blocks in which the Egyptians and Greeks had sheathed their monuments, the Romans did it cheaply and quickly using tiles. They were the proud inventors of stone cladding.

In the first century AD, when the first emperor of Rome, Augustus, went on a building spree, he replaced the city's embarrassing collection of earth and wooden temples with brick and concrete constructions boasting marble façades. As well as marble, the Romans particularly admired the travertine stone that hot volcanic springs east of Rome laid down in enormous quantities. Travertine's lightweight structure and decorative appearance made it one of ancient Rome's most popular building stones, furnishing some of the city's greatest buildings, most notably the Colosseum – but even here, behind the travertine blocks, is Roman concrete.

The Romans realised that lighter stones like tuff and travertine allowed them to save on labour costs, with no need to pay skilled masons to chip away in a marble quarry. The pre-made concrete in particular meant they could build more quickly – which for an empire expanding so fast was a valuable asset. However, there was a problem – the lightweight building stones like tuff and travertine weren't as strong and stable as those earlier building blocks. The discovery that helped the Romans overcome this and build big was not a material but a shape – the arch.

The white stone that you see all over Rome, from the fountains to the old noble palazzos and from St. Peter's church (shown here) to the Colosseum, is almost always travertine rather than marble. A soft limestone rock which grew out of calcium-rich spring waters heated up by central Italy's hot volcanic underworld, travertine's decorative appearance, with its layers, gaps and bubbles, and the huge deposits of it close to the city, ensured it was one of the most popular building stones in Imperial Rome.

ARCH RIVAL

The arch, a curved structure made from lots of small blocks of stone, was a revolutionary labour-saving architectural device. In truth, arches had been around for centuries, having been used previously in Mesopotamian, Egyptian and Greek sewage systems, but it was the Romans – having probably learned about the arch from their Italian predecessors, the Etruscans – who realised how useful they could be above ground and brought them out of the gutter.

The idea of the arch is to cut each arch stone so that it is wedge-shaped, with the widest point at the top. As a result, no stone in a continuous semicircle can slip down – each keeps its neighbours in place. However, the arch is not fully self-supporting until the last stone, the keystone, is in place. The load carried above the keystone of the arch transmits an outward thrust to its supporting columns. If such a force causes a column to bend even a little, then the arch will collapse. Therefore, the linked trains of arches that characterise Roman bridges and aqueducts were devised so that each arch supports its neighbour. To avoid the critical end arch failing, the Roman builders either anchored them against bedrock, such as rocky valley sides, or built the arches in a circular or oval plan, such as in the multi-tiered archways of the Colosseum, since this way there were no ends.

Arches and concrete combine in the magnificent centrepiece of Roman building ingenuity – the Pantheon. Built around AD 120 by Emperor Hadrian as a temple (the Pantheon literally means 'all the gods'), this giant round building has a domed roof spanning over forty metres – the largest span unsupported by columns that the ancient world had ever seen. Part of the secret of how this huge roof stayed up was that the dome acted as a

Previous page: The Romans' discovery of the arch ushered in the age of a revolutionary architectural form. Although arches were prone to collapse, the use of circular archways, most famously used here in Rome's spectacular Colisseum, were an ideal way of ensuring that individual arches were always supported by their neighbour without any loose ends.

whole series of arches that transferred the weight away on to the supporting walls instead of bearing straight down. So the weight of the dome is taken up by its outer walls, which are fronted by thin brick but filled with concrete. Different types of concrete are used in different parts of the building, and this is the real key to why it does not fall down. At the bottom, the thick foundation walls needed to have something very heavy and tough, so Hadrian's builders added lumps of basalt lava. Higher up, to keep the weight down, they used something rather lighter – volcanic tuff. Finally, in the upper sections of the dome they needed very light concrete to keep weight to an absolute minimum, so they used frothy pumice.

Thanks to their magical volcanic concrete, the ancient Romans had mastered a uniquely different and monumental architectural form that rivalled

the post-and-lintel temples of the Greeks and the triangular pyramids of the Egyptians. The arch went on literally to hold up an empire, being used to build bridges and aqueducts from Syria to France. The Pantheon was the Romans' ultimate engineering achievement – it would be well over a thousand years before anyone could better it and it still stands as the world's largest purely masonry dome. Centuries later, the builders of Gothic cathedrals would realise that arches could be as high as they wanted, providing their spans were wide enough, and high steep arches supported by narrow vertical columns would give buildings even more internal space – but Gothic arches were slow to put up, taking many decades to build a single church. The Roman ones could be thrown up quickly – much better for an empire in a hurry.

Above: The columns of the Colisseum in Rome are a façade, as is the marble-looking front. In fact, Rome's most famous landmark is clad in soft travertine slabs covering concrete below, and the multi-tiered archways bear the structural weight.

The Pantheon in Rome, the crowning achievement of Roman architecture, boasted the greatest domed roof of the ancient world. The secret to how it stayed up is found in the amazing volcanic concrete that sits behind its decorative brick façade.

TRADING PLACES

The geology on the doorstep of the main Mediterranean empires had provided the raw rocks to construct their greatest monuments and build their cities, but the grandeur of cities and the rise and fall of empires were based on wealth, and wealth came from trade. Trade is simply a way of redistributing basic resources and commodities, moving goods from where there is a surplus to where there is a need.

By and large, raw rocks were not the objects of Mediterranean-wide trade, their value being too low and their nature too bulky to be worth it, but there were exceptions. Ironically, it was arguably the highly prized volcanic obsidian that kick-started trade in the Mediterranean – people were prepared to travel long distances to the handful of isolated volcanic islands where it was found, principally Lipari in Italy's Aeolian islands and Milos in the Greek Cyclades. To get to these far-flung outposts of the ancient world required the first regular journeys across the open waters of the Mediterranean Sea. Thousands of years later, with international maritime trade flourishing across the region, marble would also achieve such cult status as a Roman building stone that it too would become a valuable tradable commodity.

In general, however, the fact that everyone had some kind of building stone on their doorstep ensured that rocks themselves were never significant objects of trade. As we shall see in the next chapter, it was a very different story for what was contained within them.

CHAPTER 4

THE AGE OF METALS

WHILE STONE IN all its variety had been the primary raw resource for weapons and tools, and for buildings, throughout human history – some two and a half million years – all this time it held within itself the very material that would finally topple its supremacy. As our ancestors worked the stone, learning more and more about its properties and possibilities, they would have noticed the effects of weathering, which leached minerals from it. These minerals would first give them the means to express themselves artistically, and then lead them on to the next all-conquering technological marvel: metals.

PRIMARY COLOURS

Fifteen thousand years after their creation, the paintings in the Lascaux caves in south-western France are a spectacular window into a prehistoric world. Hundreds of images of animals – mostly horses and bison, with some ibex, stags and mammoths – are brought to vibrant, immediate life on the walls. Along with the Altamira cave in northern Spain, Lascaux represents the crowning achievement of prehistoric art in Europe, carried out by the Magdalenian people – and the materials for their expression were drawn mainly from the rocks all around, the rocks that for so long had been a purely practical resource.

Just as early tool-makers and builders used what they had to hand, so prehistoric artists used what was on their doorstep. They obtained black from carbon created by burning wood to make charcoal; white came from ground-up chalk – which is basically the crushed limestone made from the remains of skeletons of coccoliths, tiny planktonic creatures, and which was a common rock type in many parts of the Mediterranean – but mostly, they painted with the residues left behind when rocks rot.

Rocks rot when they are exposed at the Earth's surface to the elements. Carbon dioxide in the atmosphere combines with water to form a slightly acidic rainfall that starts to break down the minerals chemically within the rock. Minerals are collections of chemical elements organised in unique arrangements, rather as letters make up words – and just as words get arranged into sentences, different sets of minerals put together in certain ways result in rocks. Regularly occurring combinations of minerals are the 'grammar' that gives us the main rock types discussed in the last chapter: the limestones, sandstones, marbles, granites and basalts. Although more than four thousand minerals are known, only about thirty are common at the Earth's surface, and it is these that make up the bulk of the rocks we see.

Different minerals rot, or weather, to different degrees. Quartz, or silica, a

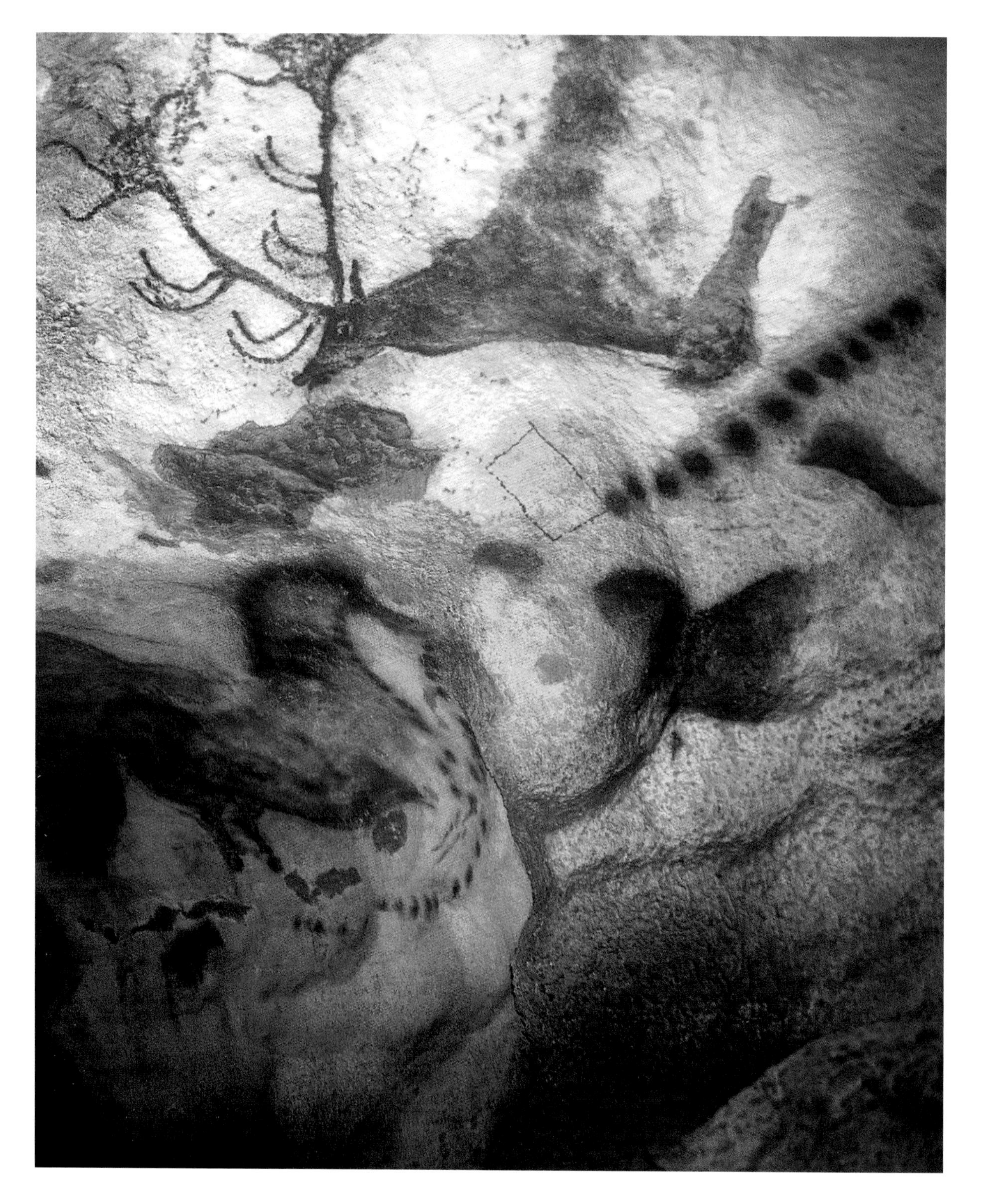

pairing of the two most abundant elements in the Earth's crust (oxygen and silicon), is so stable that it is virtually immune to chemical attack. In contrast, most common minerals decompose into flakes of clay particles when exposed to air and moisture. As minerals break down they release their elements, some being dissolved and carried away by water, while others less soluble are left behind as a clay-rich residue that reacts with the oxygen in the air to form oxides. After silicon and oxygen, the most common elements in the crust are iron, aluminium and manganese. These elements are hard to flush away and readily form oxides at the surface. Weathering of most rocks will produce iron oxide, or hematite, a reddish brown material that forms ochre, an iron-rich clay powder. Other iron–oxygen combinations give minerals with other hues: limonite gave our ancestors yellows and oranges, while magnetite provided browns. If they wanted a darker brown substance then another weathering residue was often available to hand – manganese oxide.

Previous page: The weathering of rocks provided prehistoric artists with a natural palette of colours right underneath their feet. With exposure to the elements, rocks rot away to leave behind a residue of iron-oxide minerals, which here at Roussillon in Provence turns sandstones cliffs into a wonderland of red, orange and yellow ochre.

Opposite: Some of the world's earliest paintings adorn the walls and ceilings of Lascaux cave in France's Dordogne region. Painted by Stone Age people at the end of the last Ice Age, this natural art gallery depicts life in the freezer using pigments that our ancestors literally scarped from the ground.

With charcoal and chalk, these colourful weathering crusts gave early humans the pigments for the prehistoric paint palette. All they had to do was scrape it up from the ground, grind it into a dust, mix it with some water, oil, blood or urine and then paint it on. The early pigment hunters would have been experts at spotting the particular plants that thrived on the brownish and reddish metal-enriched soils, but often, scraping beneath the surface of dull oxide crust, the prospectors in some places found far more colourful rocks – dazzling greens, brilliant blues and vibrant oranges. What made these rocks so colourful in the first place?

Colour is more than skin deep; in fact it's made in the atomic structure of elements. In most elements, the atoms are so arranged that they share electrons with each other. Each carbon atom, for example, has eight electrons – four of its own and four that it shares with neighbouring atoms; in diamond, every carbon atom links each shared electron to that of another carbon atom to form a stable and very strong three-dimensional framework. (Graphite too is made of eight carbon atoms but these are linked together horizontally as sheets, with only occasional vertical links between the sheets, so that the layers are able to slide easily over each other, making graphite soft and slippery.) Some elements have a careless tendency to lose electrons, and this ends up in electrons roaming freely in a 'sea' through the atomic structure, casually attaching themselves to atoms by so-called 'metallic bonds'. These elements are called 'metallic', and the compounds they make with other elements are what is meant by metals.

It is the roving electrons that give metallic elements their colours. When sunlight falls on to a substance, its electrons absorb the energy the light contains. A shaft of daylight looks white but in fact it is actually a rainbow of colours, each colour being a different wavelength or 'energy' of light. Electrons are picky eaters; they selectively absorb particular wavelengths of energy – the light in the parts of the rainbow that they are partial to – but reflect back those light energies or wavelengths that they have no 'appetite' for. A substance whose electron population has a penchant for blue, green, yellow and orange light waves will devour these but reflect back the unwanted red light waves, making the material appear red. In other words, the colour we see in a substance is actually the colour it isn't! It is the light waves that a material can't absorb that our eyes detect as colour. Metallic elements with lots of roving electrons are especially selective which, because they reflect much of the light waves they receive, means they appear especially bright and colourful.

Of course, the early prospectors had no knowledge of elements, atoms and the like. As they searched for ever more colourful pigments they simply found that in some places the rather dull carpet of red and brown oxides hid a gaudy interior of bright hues. Green 'malachite', blue 'azurite', brassy 'chalcopyrite', orange 'realgar' and yellow 'orpiment' would gradually give early painters a whole new palette of stunning pigments. More impressive than their colour, however, was that many of these rocks sparkled. Much of the natural glitter came from pure metallic elements – those elements stable enough to be found on their own rather than as mixed metal compounds. Such 'native' metals – gold, silver and copper – were the bright shiny things that ushered in a new age that would gradually see the end of the Stone Age.

■ METAL DETECTORS

Metals seem to have been discovered by early Neolithic peoples farming in the Anatolian hills of south-eastern Turkey around nine thousand years ago. The first metalworking was a small cottage industry, exploited by scattered villages lying close to the most obvious metal-rich seams. That arrangement would change with major social developments happening down on the farming plains of Mesopotamia – a crucial phase in human history that is the subject of the next chapter. Large numbers of people needed to be brought together in ever-larger villages and towns to develop and maintain cultivation and store the growing food surpluses. As towns got larger, more crops could be harvested. From the extra wealth came a demand for a much wider and more varied range of raw materials – building and bonding materials, ceramics, glass, colouring pigments and textile fibres. Social wealth began to support individuals and classes that did not participate directly in food production. Besides priests and nobles, there were new members of society who did specialist non-agricultural work. Among them were the artisans who worked with the most expensive and most difficult to find of all raw materials – metals.

Metals were a resource that could be properly exploited only once complex communities and societies had come into existence. Unlike working with stone, where a craftsman produced a desired tool or weapon by hewing his raw material on the site where it was extracted, making metal implements involved several stages and different groups of people. Prospectors discovered the rocks that contained the right metallic compounds, miners dug it out, and founders melted the metal in kilns or crucibles and cast it in portable terracotta moulds. These raw metal ingots then passed into the hands of merchants who carried them on long journeys to their destination, where a second cycle of work began with the metalsmith, who refined, remelted and recast the raw metal. Then another artisan entered upon the scene to hammer it, weld it and make it into a metal plate that might be embossed or engraved, or transformed it into wires. Because metalworking was so difficult, requiring precise control over a number of variables and following a carefully worked out set of steps, it soon became an area of important and rapid technological innovation.

Early metalworking, though, had a slow start. Few of the metallic elements were available in pure, or native, form and these – mainly gold, silver and copper – were too soft to be of real practical use. Not strong or hard enough to replace stone for tools or weapons, the materials that would eventually out-perform stone were initially more for fun than function. They were used to satisfy a growing demand from the newly rich members of those expanding settlements for eye-catching ornaments and jewellery – pins, pendants and beads – objects that were conspicuously costly.

Gold was particularly prized because in its native form it could be gathered and worked without difficulty. As one of the softest of metals, it could be hammered into shape easily. This meant it could be beaten out into thin foil sheets without the need to reheat it so as to prevent it cracking. The first fabulous gold jewellery and artefacts appear in the Near East around 5000 BC, and two thousand years later gold objects were being made and traded throughout the Mediterranean.

The early metalworkers discovered that hammering soft copper hardened it, while heating it again prevented it from cracking or splitting, something that made it even more durable. They began to experiment with different ways to manipulate it. Around 5000 BC they found that if they heated galena – lead sulphide, the most common lead mineral – in their basic kilns, pure lead dripped out and could be cast in terracotta moulds. The practice was soon applied to copper-oxide rocks with surprising results: although requiring hotter kilns than for lead, and although the molten copper that was extracted had impurities in it, the metal that resulted was often harder and more durable than the native copper.

Gradually, the first useful metal implements appeared, such as natural copper hatchets that turned out to be better than those made of flaked stone. This was the start of the slow realisation that in spite of the long labour process, metals offered what no stone could: inexhaustible plasticity combined with superior hardness and resistance. Over the next few thousand years – the Copper or Chalcolithic Age – not only would every existing implement eventually be made better in metal, but new ones could be invented in a huge variety of forms and for a huge variety of uses.

■ MOUNTAINS OF METAL

Demand for metals was growing, but the problem was that the raw material was found in only a few places. These places were not on the densely settled agricultural lands of Egypt and Mesopotamia, but were instead scattered across the mountain ranges of the Mediterranean. That is because mountains are the places where rocks and minerals have been brought up from great depths, and it is from deep within Earth's crust, in many cases deep within the mantle of the planet's interior, where metals are formed. The dense metallic compounds settled here when the early Earth separated out from its gassy stew of elements, but circulating plumes of hot mantle rocks brought metals up to the crust, moving very slowly – just a few centimetres a year. This magma of melted rock and dissolved metals oozed up along cracks and fissures, driving a jet of gas ahead of it to force open new pathways. In the superheated steam was a consommé of lighter elements, like silica, and chemically unreactive metals such as gold; following behind in the magma were the heavier metal deposits, typically of copper, iron and zinc. A few kilometres from the ground surface, where the pressure of the overlying rocks was lessened, the head of boiling steam blasted upwards, eventually cooling to clog up near-surface faults and fractures as gold-rich silica (quartz) veins. Deprived of its advance jet, the magma and its metal cargo were left stranded at greater depth.

That's not the end of the story. Cold groundwaters that trickled down from rainfall at the surface now passed through these hot metal deposits and, invigorated with heat, rose to erupt at the surface as hot springs and geysers. Anyone who has visited, or bathed in, these thermal waters will know the 'rotten egg' smell that often clings to them, an indication that hydrogen sul-

The mountains of the Mediterranean are like a giant metals factory. Here in the Jordanian highlands, a zigzagging network of greenish veins of copper-rich rock are exposed in the walls of a large wadi. The discovery of such colourful rocks in the eastern Mediterranean mountains almost ten thousand years ago lead to the start of metal-working which in turn ushered in the Bronze Age

Much of ancient Egypt's mineral wealth lay here in the mountains of the Eastern Desert, where cracks and faults in the crust caused by the opening of the Red Sea allowed metal-rich magmas and fluids to ooze up from deep within Earth towards the surface, where the metallic cargo were concentrated into mineral seams like this ridge of basalt rock.

phide (along with carbon dioxide) is escaping with the water. At depth, these gases became dissolved in the groundwater and made it extremely acidic; this allowed it to act like a scavenger, collecting and dissolving metallic elements it encountered in the rocks as it passed through. The leached metals were carried upwards where the water cooled, metallic elements combining with the sulphur to form sulphides and with the carbon dioxide to form carbonates. As a result, what were once scattered metallic specks became naturally concentrated into 'ores'.

Over millions of years, as growing mountains pushed up rocks and erosion stripped off the overlying strata, the deep ore-rich zone became exposed at the surface. Rocks, faults and fractures that had been stuffed with metals many kilometres below the ground were now just beneath the feet of early prospectors. The mountains of western Iran, the Caucasus, Armenia, Anatolian Turkey and Cyprus are home to massive copper sulphide deposits. Further west, rich lead, silver and gold deposits occur in the ancient metamorphic mountain core of the Aegean, from western Turkey across to eastern and northern Greece. The Alps and the Pyrenees too offer rich seams of metal, as do the mountain ranges of North Africa. In the eastern

Mediterranean, the great splits in the crust along the Red Sea and the Dead Sea make the uplands of eastern Egypt, the Sinai and Jordan home to substantial reserves of copper and gold. Older episodes of mountain building from long before the time of Tethys form enormous quantities of sulphides in southern Spain and Portugal and southern Egypt. The Mediterranean, it seems, is made of mountains of metals.

Along with their riches in metals, many of the mountains bear precious stones as well. Most gemstones are minerals that form in the crust or upper mantle alongside metals, and they are brought to the surface along similar pathways. In fact, it is the metallic elements acting as impurities in the gems that give most their colour. Emerald, garnet, jade, turquoise and amethyst were all mined in antiquity in the eastern Mediterranean, mostly by the ancient Egyptians. However, they had to import the widely used blue stone lapis lazuli from its natural home in Iran and Afghanistan. In many areas, ancient hot fluids had stripped silica out of the rocks too and left it behind as veins of coloured agate and opal. Gem-quality opal was highly prized by the Greeks and Romans, but unfortunately for them this was incredibly rare in the Mediterranean; it results from hundreds of millions of years of tropical weathering, and today Australia provides almost all the world's supply. And the greatest prize of all eluded the Mediterranean – diamonds being found only in the ancient interiors of the great landmasses such as Australia and southern Africa.

MEN OF BRONZE

In the seventh and sixth millennia BC, much of the early metal, especially copper, for Mesopotamia came from mines deep in its mountainous northern border. The great lumps of native metal could simply be loaded on to rafts or hide boats and floated down the Tigris or Euphrates rivers to reach the heart of the empire. Mesopotamia jealously guarded its mountain mines and, together with its superpower rival Egypt, managed to maintain a close monopoly on the trade in metal. All that would change when the most obvious sources of the easy-to-get copper became exhausted.

By around 4000 BC, prospectors in many areas were being forced to hunt for metals in more unlikely guises. In their search, they came upon richly coloured arsenic-bearing copper oxide ores, which in turn led them to the metal sulphide deposits below. These were far more widespread than the scattered pockets of native metal but their complicated chemical make-up created difficulties in getting the raw metals out. Since they had already realised that impure sources of copper produced a finished product that was stronger than others made from native copper, they began experimenting with smelting these arsenic-rich rocks to produce copper alloys that were far harder and more durable than pure copper. As well as arsenic, these earliest alloys contained appreciable quantities of lead, nickel, antimony and occasionally even silver or tin, and it was a long road for most cultures to turn these complicated ingredients into implements with precisely the right properties. Trial and error showed that additions of lead and antimony allowed the copper to pour more easily, but it was the addition of tin that revolutionised the copper world.

Tin is a rare native element generally oxidised to form the mineral cassiterite, giving a shiny rock that the ancients confused as a form of lead and so attempted to fuse with copper. Amazingly, adding 10 per cent tin to copper produced a new metal, bronze, that was harder and more resistant than any other copper alloy, and hence was far more useful. Quickly the scattered deposits of tin in the Near East were intensively exploited and soon exhausted. As a mineral most associated with granite magmas, much of the Mediterranean tin was still locked away many kilometres down in the crust. Much older tin deposits were found in the ancient eroded mountains of Cantabrian Spain, Brittany, and Cornwall in western England. To obtain the essential ingredient required long and perilous sea journeys from one end of the Mediterranean Sea to beyond the limits of the other. If there was ever one material that could be said to be the catalyst for international maritime trade in the Mediterranean and beyond, it was tin.

The superiority of the new bronze ensured that the centres of metal-working power shifted from where the raw rock was extracted to the various staging posts where the raw materials needed for the increasingly complex metal processing operations could be brought together most easily. Islands like Cyprus, which was able to combine local wood fuel and copper ore with tin imported from abroad, became flourishing hubs of maritime trade. Elsewhere trade combined with the mineral wealth on their doorstep allowed local empires to emerge – Hittites in Turkey and Minoans in Crete – that challenged and checked the great Egyptian and Mesopotamian metal monopolies.

The Bronze Age began in the Middle East around 2500 BC when copper and its alloys were used for weapons and ornaments, though stone tools and implements were still being made. It was perhaps inevitable that the rise of a new metal that gave weapons far superior to anything that had gone before would prompt the first major clashes of peoples, initiating an era of invasions, wars, pillage and political crises –and it appears that what began with the sword, also ended with it, as the Bronze Age dissolved in violent turmoil around 1200 BC. This was the time of widespread destructions and migrations attributed to the sweeping invasions of the so-called Sea Peoples, warriors who may have been Dorians from northern Greece or Philistines from Palestine. It culminated in the annihilation of virtually all of the major eastern Mediterranean Bronze Age civilisations and the rearrangement of whole populations. The Hittite empire in Anatolia, the North Syrian centres of Ugarit and Alakh, the Mycenean civilisation of Greece (which had superseded the Minoans), and a number of important Cypriot city-states were all casualties of this great upheaval. Egypt claimed a pyrrhic victory over the Sea Peoples when they reached their shores, but in fact the wind had gone out of the Egyptian empire too and the days of its New Kingdom were numbered. As one flourishing centre after another fell, the cosmopolitan Bronze Age worlds became increasingly fragmented and isolated. The so-called Mediterranean Dark Ages had arrived.

Quite why the Bronze Age empires never recovered from these invasions is a thorny issue to which we will return in later chapters, but the disruption to the bronze trade is probably one part of the answer. It wasn't so much that copper itself became unavailable – there were abundant deposits in the region and certainly enough to keep local needs going – but since tin ores were exhausted in the eastern Mediterranean, a shortage of foreign tin of any sort would have been disastrous for the bronze. It seems likely that the breakdown in maritime trade caused by the sweep of the Sea Peoples caused the eastern Mediterranean to be cut off from the regular supply of western tin that it had enjoyed for the previous thousand years. A lessening of the tin content of bronze tools or weapons would greatly reduce their strength and durability. Without an obvious replacement to hand, the region was forced to revert to more inferior materials. One of the metals that becomes increasingly common after the fall of bronze was one that within a few centuries would come to surpass it – iron.

■ AN IRON HAND

Given its long-standing strength, hardness, availability and cheapness, iron would seem to have had many obvious advantages over bronze, and had clearly been long known in the form of the iron oxides discovered by the early pigment hunters, but the adoption of this supposed new supermetal during the Dark Ages was surprisingly slow. The almost reluctant emergence of iron is a complicated story that derives much from the fact that it is not like most of the other metals that circulated in the Bronze Age Mediterranean.

It may be the most abundant metal on the planet, but most iron is locked away deep within its interior. In the Mediterranean, some iron occurs as a minor impurity in copper deposits, but mostly it is found in laterites, iron-rich red soils many metres thick. These laterites were formed by the rusting of the land surface around a hundred or so million years ago when the shores of the Tethys enjoyed warm and wet sub-tropical climate and the uppermost rocks became deeply weathered. Although now buried by younger strata, such deposits of insoluble iron (and aluminium) oxides are abundant across the Mediterranean, and most regions have them on their doorstep.

Because iron was formed differently, metalworkers found it behaved differently in their kilns. For thousands of years, copper and then bronze had been processed by melting varieties of coloured stone in a wood-fuelled oven to produce a flow of metal. However, if those stones were iron minerals – hematite, magnetite, limonite – the process didn't work. Instead of molten rock, only a spongy, pockmarked mass was produced, and the soft globules of iron inside quickly solidified due to the high melting point of iron (1500°C) and remained hidden. Since the small ovens could never reach the melting point of the iron ore, the ancient metalworkers never saw cast iron, though they did realise that the spongy lumps could be heated to red-hot and then hammered down to give a reasonable piece of metal.

Historians long assumed that iron ought to have been superior to other metals in antiquity because it is known today to be so hard and durable, but in fact, while pure iron is harder than pure copper, it is less hard than hammered copper and certainly far softer than bronze. Although iron can be strengthened by the addition of carbon to make steel, this would still not make it superior to hammered bronze. It is only when steel is cooled in water (quenched) that its strength becomes vastly superior to that of bronze. Thus unless iron is 'steeled' by the addition of carbon, hardened by quenching and then heated to reduce brittleness, it does not have the hardness, strength and ability to take a cutting edge as bronze did. Although none of these processes is difficult on its own, they all differ considerably from those practices that would have been familiar to bronzesmiths.

The early production of iron was a pretty hit-or-miss affair. Studies of iron artefacts dated to around the eleventh century BC show that sometimes iron tools or weapons were of high quality and sometimes they were virtually useless. What was lacking was a precise production line with a set of exact procedures and careful temperature control, since this was something that copper and the other metals worked during the Bronze Age simply didn't need. Iron was indeed potentially far superior to bronze in many respects, but it seems likely that it took the centuries of privation in the Dark Ages for people to realise how to make it that way. Iron was exploited at first only with reluctance precisely because of its unsatisfactory and unreliable properties, and where possible bronze continued to be used, albeit sparingly, but the general shortage of bronze and the overwhelming abundance of iron forced people to persevere experimenting with this tricky metal.

With unreliable or reduced access to tin, that critical raw material, the peoples of the Eastern Mediterranean had little choice but to turn to the iron on their doorstep and make what use of it they could. In many respects the Dark Ages mark the quiet spreading of ironworking knowledge around the eastern Mediterranean. The earliest specialists in iron had been the Hittites in Turkey but, with their overthrow by the Sea Peoples, wandering bands of metalworkers carried iron's tricky secrets south and eastwards. Because local peoples already had iron deposits nearby that they had been struggling to work with, iron technology spread faster than copper or bronze. It soon reached the iron-rich territories of Greece and the Levant, and after only a few centuries had passed through the ironfields of southern Italy and Umbria to reach central Europe.

Across the region, a new age – the Iron Age – had begun. Small agricultural backwaters digging their own small deposits of iron ore were able to produce enough of the metal to free themselves from their dependence on those who controlled the vast trade routes of copper and other precious metals. From the ironsmith forges now came highly resistant and cheap farming tools. Areas like Greece flourished agriculturally, and to maintain food supplies to their growing cities they spread outwards to found colonies in Italy and, more importantly, Asia Minor (modern Turkey). New trade centres blossomed and the Aegean once again found itself at the hub of an international trade network that plied the Mediterranean and nearby seas. An increase in trade meant an increase in wealth, and led to the birth of money.

COINING IT

From the earliest movements of trade, people needed to be able to exchange one commodity with another. In many cultures, common units of value were used to make that exchange easier. In some the standard currency was cattle, though it was rather cumbersome to carry around. The use of such live money continued until recently. Well into the twentieth century, the Kirghiz peoples of the Russian steppes used horses as their main monetary unit, with sheep as a subsidiary unit; small change was given in lambskins. Ancient societies similarly created currency out of essential everyday objects, such as spades, hoes and knives, but by the eighth century BC, the hub of

The Parthenon temple on the sacred limestone hill of the Acropolis in Athens, Greece, built in the 5th century BC with the new wealth from the cities nearby silver mines and clothed in local Penteli marble

international trade in the known world centred on the Aegean coast of western Turkey – Ionia. Here, pressured by the rapidity and frequency of exchanges, the Greek merchants of Ionia began the custom of doing business using precious metals cast into small ingots – and the most obvious metal to use was the one in their backyard.

The mountain hinterland of Anatolian Turkey was rich with gold, much locked away in vast sulphide deposits. Many ancient Bronze Age cities in the region established gold mines and grew rich on their proceeds. One was Gordian, home to legendary King Minos, whose cursed affliction that everything he touched turned to gold was only lifted by him washing his hands in a river which thereafter flowed with gold. In fact, most of the Anatolian rivers really did flow with gold –so much so that at Sardis, capital of ancient Lydia in western Turkey, people draped sheepskins in the river to catch the tiny specks of gold washed down from the nearby mountains (a practice that may well be the origin of the Golden Fleece, which Jason famously tracked to the mountains of north-eastern Turkey). These golden waters arise because as the gold-bearing rocks weather and break up, the flecks of this chemically stable element remain unweathered and wash into the rivers. Because gold is soft or malleable, when the flecks tumble about in running water they are welded together by bumping and banging, forming nuggets that become too heavy for the water to carry and so are dumped in sediments in the upper reaches of rivers. Early Bronze Age prospectors fruitlessly searching for tin-bearing cassiterite in mountain streams found instead a bonanza of gold, but after this early goldrush when the largest and easiest gold nuggets were collected, gold became harder to find. Not only that but most of the gold that was found had some silver mixed in with it; if there is more than 20 per cent silver, the alloy is called electrum.

Electrum, although far less valuable than gold, became widely used by the Ionians for their ingots of currency, but in 700 BC, the Lydians of Sardis had a breakthrough. They discovered how to separate the gold and silver in electrum, and they set up a refinery for doing it. It was a simple procedure. The electrum was placed in a vessel packed with table salt (sodium chloride) and sand and then heated up in a furnace to allow the silver to react with the chlorine in the salt to produce silver chloride, this compound moving from the electrum across into the sand. The unreactive gold could then be removed, and the silver later collected by a different purification method.

Having discovered the way to make electrum give up its gold and silver, the Lydian kings replaced the ingots of electrum with gold ingots of a fixed weight, guaranteed by the royal authority whose lion's head seal was stamped into the ingot. Later, when metallurgical skills improved and these ingots

became more regular in form and weight, the seals served as a symbol of both purity and weight. Money had been invented.

The first real coins were probably minted some time in the period 640–630 BC and quickly gave the merchants of Ionia many advantages. First, money drastically reduced the material volume of transactions, making trade easier and smoother. Also, the basic currency could be divided into multiples (not an easy thing to do with a cow), so that any value could have a corresponding material weight. It replaced cumbersome iron, lead and electrum ingots that were all easy to counterfeit and, being of lower intrinsic worth, were inconvenient for expensive purchases.

The Golden Age of ancient Greece had begun. The invention of the Ionian merchants and Lydian kings soon spread to all Greek cities and then around the eastern Mediterranean. On the back of this invention, the Ionian traders derived great profits from a soaring increase in international trade. Small and medium farmers, tied to traditional bartering, grew weaker, while new classes less directly tied to the land arose. These traders, money-changers, money-lenders and bankers no longer exchanged their services for goods in kind, but for money. Gold itself became a form of wealth, and soon would become *the* form of wealth.

Liberated from gold, silver too became valued for coinage, and areas with large deposits of silver flourished. Unlike gold, silver is a far more gregarious metallic element, readily attaching itself to other metal compounds. One of its favourite partners was lead, but as the Bronze Age metalworkers had discovered four thousand years earlier, lead is relatively easy to extract from its ores and so could be encouraged to give its silver up, typically releasing one part silver for every 300 parts lead. Air was blasted over a melted mixture of lead, silver and other metal impurities (gold, zinc, copper, arsenic) until the lead oxide that forms with the air absorbs the metal impurities and leaves the silver and gold – and of course the Lydians had shown how then to extract the gold.

In the seventh century BC the muddy plains of Attica, weathered down from the metamorphic hills that surrounded its capital Athens, blessed the people with excellent and generous supplies of potter's clay and made Athenian pottery the best in Greece, but another even more valuable source of wealth lay to the south-east of the city – the lead–silver mines of Lavrion. The Lavrion deposits were remarkably rich – the ore contained up to 50 per cent galena – and the lead smelted from this ore had 0.1–0.4 per cent silver. The lead itself was used for anchoring iron clamps, covering the hulls of ships and for making sinkers for fishing lines and standard weights; it was also added to bronze to make the molten metal more fluid and easy to cast; but it was its role in liberating silver and gold that was most valuable. It is estimated that between one and two million tonnes of lead and 8,000 tonnes of silver were produced in these mines in antiquity.

Although silver had been mined at Lavrion since late Bronze Age times, Athens began to grow steadily rich on the mines, around the sixth century BC. With the new demand for coinage spreading across from the other side of the Aegean, it issued its first set of silver coins, known as 'owls'. Then, the city's revenues from the mines dramatically increased when a deeper and far richer galena vein was found early in the fifth century BC. One of the Athenian leaders at this time, Themistokles, persuaded the citizens of the city to use this windfall to build a fleet of warships, an act that quickly made the city a naval power. In 456 BC Athens took on and defeated the finest naval fleet of neighbouring Aegina, forcing the island to accept Athenian 'owls' and to stop minting its own 'turtle' coinage. Coinage, after all, was first and foremost a civic emblem – to strike coins with the badge of the city was to proclaim one's political independence. A few years later, Athens issued an edict ordering all 'foreign' coins to be handed in and compelling all its allies to use the Athenian standard of weights, measures and money. With the wider Greek world bullied by Athens into accepting its silver coinage as standard currency, the later conquests of Alexander the Great would take that monetary uniformity over much of the known world. In the days before mechanical printing was invented, coins were by far the best propaganda weapon available for advertising Greek or any other subsequent self-respecting civilisation.

■ LEAD WEIGHT

In the frozen wastes of Greenland, ice layers dating back thousands of years retain a remarkable record of the rise in coinage in the Mediterranean. Lead particles measured in the annual ice layers are present at barely detectable levels in prehistory, but levels start to rise at 8,000 years ago when Mesopotamians began using lead-smelting furnaces, and they soar dramatically around 600 BC when the Greeks started to conduct lead-mining operations on a massive scale.

The Greenland ice cores reveal that lead concentrations in the atmosphere peaked around the second half of the fourth century BC when the Greek silver mines were at maximum output (probably no more than 10,000 tons per year) and before slave revolts forced them into centuries of decline. Then, after conquests in Spain and Greece gave the Romans access to the silver mines of its neighbours, lead levels were on the rise again. They peaked around the end of the first century BC, when lead production across the Roman empire was estimated to be 100,000 tonnes per year. Ever practical, the Romans rarely opened up new areas of mining, but simply took over the operation of existing ones and expanded production, relying heavily on enormous workforces of slaves, prisoners of war and convicts. The Roman lead mines near Cartagena in Spain, for instance, required some 40,000 slaves, who had to be constantly replenished because of the appalling work conditions. It wasn't just that mining a mountain was tough work – it was that the lead itself was deadly.

Aside from smoke, lead is probably the oldest man-made atmospheric toxin. Although we're now used to hearing about lead poisoning from petrol and the benefits of lead-free fuel, mining for metals has been polluting our planet for over two thousand years. The virtual absence of free lead in our biosphere before the earliest metal-working societies is presumably why human beings, along with other terrestrial life forms, evolved no biochemical defences against the element. By contrast, our bodies possess very effective mechanisms for limiting the accumulation of other potentially dangerous but naturally abundant metals such as copper and iron. Lead atoms have a great affinity for sulphur atoms; hence the most widespread mineral of lead is lead sulphide, or galena. Lead's attraction to sulphur is also responsible for its toxicity. If lead is introduced – even at levels as low as a few parts per million – its atoms will combine with sulphur in the proteins within our bodies, disrupting them and causing a wide range of enzymes to malfunction.

Lead is a cumulative poison: its most serious effects do not appear until a critical amount, in the order of a few grams, has entered the body. After lead is ingested, it appears first in the blood, but is soon distributed throughout the internal organs and the skeleton, where it may be retained for decades. Our bodies can't eliminate it as a waste product, so it accumulates in the bones and tissues and clings to the red blood cells. Depending on how much they take in, those unfortunates who have ingested lead suffer a wide range of symptoms: headache, insomnia, jaundice and diarrhoea to start with, then severe stomach pains, gout and extreme, even complete, constipation caused by paralysis of the intestinal tract. Finally come serious central nervous system disorders – deafness, blindness and paralysis – as well as imbecility, hyperactivity, dementia and, of course, death. Even if lead doesn't kill, in smaller doses it causes sterility in men and infertility in women. It can also lead to an increased likelihood of stillbirth and miscarriage. Lead is not a human-friendly metal.

Unfortunately for the Romans, they were infatuated with the stuff. Even though most of the Roman world lived thousands of miles from the dusty lead-filled smog of their nearest lead mine, the metal was fantastically useful in Imperial Rome. It was readily available, inexpensive, easy to mine, smelt and work, and very resistant to corrosion. The Romans used it to line their aqueducts, used it in mortar in stone structures, and most extensively used it in plumbing. So, when the Romans constructed their extraordinarily sophisticated public and private baths, lead was the metal of choice for the miles of water piping. Fortunately, the sheer volumes of fresh water flowing through the pipes and gushing from the city's fountains and basins probably prevented ordinary citizens from ingesting too much lead.

The same might not be said for those enjoying the good life. Upper-class Romans were inordinately keen on their food and drink, to such an extent that they became notorious for their gluttony and drunkenness, but much of that food and drink was laden with lead. To stop their wines turning into vinegar during shipment, ancient merchants resorted to a variety of preservatives. While the Greeks had added pine resin to their wines, a custom that continues today in Greek retsina wines, the Romans preferred a preservative called 'sapa'. They prepared sapa by boiling unfermented grape juice in a lead kettle until it was reduced to about a third of its original volume. During boiling, however, the grape juice leached significant amounts of lead from the kettle walls: the dark, sweet, aromatic syrup was a highly toxic concoction.

Sapa wasn't just an effective preservative for wine, it was also used as a sweetening agent in many Roman dishes. Nearly 20 per cent of the dishes in a typical ancient Roman recipe book have sapa as an ingredient. Lead would also have been present in cups and plates as well as pots and pans. Sapa contained about 1 gram of lead per litre, and wines to which it was added in the recommended proportions had a lead content of about 20 mg per litre. In Imperial Rome, the ruling classes drank hugely – apparently anything from one and a half to five litres a day. With much of the wine likely to have been contaminated by the deadly additive, anyone consuming that much would have serious symptoms of lead poisoning within weeks.

Some scientists have suggested that lead might have contributed to the start of the slow and complicated demise of the Roman empire. From the first century AD, the Roman upper classes apparently suffered from an inability to reproduce – a prominent symptom of lead poisoning. They began to die out with extreme rapidity, each generation being perhaps a quarter of the previous one. The health and mental capacity of the emperors too seemed to decline during that period, as a series of successors to Augustus succumbed to illness and erratic behaviour. Tiberius (AD 14–37) was rumoured to have been a schizophrenic, whose excessive drinking earned

Opposite: *The small Spanish town of Almaden (right) boasted the largest mercury mine in the world until a few years ago when environmental concerns over the toxic effects of the mercury vapours forced the mine to close. Today the mine buildings on the left have been transformed into a museum.*

him the nickname Biberius Caldius Mero – 'drink-loving and hot with wine'. Caligula, or Gaius (AD 37–41), a chronic alcoholic who reputedly married his sister, made his horse a consul, turned his palace into a brothel and killed innocent citizens on a whim. Next was Claudius (AD 41–54), hampered by a limp, trembling and a speech defect (some attribute this to cerebral palsy) and by continual illness. Finally there was Nero (AD 54–68), who killed his mother and wife and apparently fiddled while the Eternal City burned. Interestingly, the ice cores indicate that lead levels in the atmosphere begin to plummet around the end of the first century AD as the Roman lead mines became exhausted and closed. By then, perhaps the damage had been done to the Imperial Roman world.

■ GOOD ROCK, BAD ROCK

Regardless of the extent of lead's role in the demise of the Roman empire, geologists now appreciate that as well as providing shelter and vital raw resources, rocks can kill. In ancient times metals like mercury and arsenic were famed for their poisonous potential. Both metals were mined extensively in Roman times but today mining of these hazardous rocks is increasingly governed by environmental concerns. Until recently the small Spanish town of Almaden (Moorish Al' Maden, 'the mine'), boasted the largest mercury mine in the world, providing a third of the global supply of the metal. The Almaden mine had operated since Roman times, but such was the concern over the progressive poisoning of miners by the mercury vapours that later legislation required workers to be exposed for only a maximum of six hours in eight days per month; a few years ago production stopped entirely and the mine was turned into a museum. Across the world, the environmental costs of mining many toxic metals can be as high as the economic ones.

Lead, mercury and arsenic, together with cadmium, are recognised today to be the most lethal metallic pollutants, but others that can also be toxic are aluminium, chromium, copper, molybdenum, nickel and zinc; problems are also encountered with rare elements like antimony, selenium, thallium and silver. The health effects of these environmental pollutants are complex and varied, but cadmium is known to affect the kidneys and skeleton, while mercury and aluminium disrupt the nervous system and arsenic the heart.

Metals may be harmful, but we can't live without them. Although humans are 99 per cent oxygen, carbon and hydrogen with a bit of nitrogen, calcium and phosphorus, the remaining 1 per cent of us is made up of a huge variety of trace elements. The trace elements most essential for our health are

chromium, cobalt, fluorine, iodine, iron, manganese, molybdenum, selenium and zinc. Many of these are the same elements that can be poisonous! Elements that in small quantities do us good, in larger amounts are harmful. Too much of the main elements like calcium doesn't present any problem – you can't O/D on calcium – but toxic elements like mercury and lead are clearly bad. Elements such as copper or molybdenum are tricky – we have to have them in our bodies to be healthy, but too much and they are toxic.

The problem is that just as different types of rocks are unevenly spread around regions like the Mediterranean, so trace elements are not uniformly distributed either. In other words, geology gives different regions different health effects, and possibly different diseases. For example, it is globally recognised that low concentrations of iodine are found in soils at high elevations and in limestone terrains, both characteristics common in the Mediterranean. Medical studies show that iodine is an essential nutrient and a human being normally contains 15–20 grams of it, mostly in the thyroid gland. Deficiency in iodine can lead to serious enlargement of the thyroid gland – the endemic disease goitre – an affliction already recognised in parts of Greece and Egypt where iodine levels in drinking water are abnormally low.

While too little iodine in drinking water is a problem, too much fluorine in it is equally bad. Fluorine is an essential element and our principal source for it is water where the natural content is less than 1 part per million. A deficiency in it leads to increased tooth decay, especially in children, and so in many places it is regularly added to our water. It is also believed that taking small daily amounts of a few parts per million of fluorine over a lifetime could help to stave off osteoporosis. However, in some areas, groundwaters derived from rocks, volcanic gases and mineral springs with high fluorine levels can lead to very high fluorine levels in drinking water. High natural concentrations of fluorine have been found to lead to over-abundant deposition of the calcium phosphate which builds our bones. Too much fluorine in our bodies can be lethal. Fortunately with fluorine the difference between a beneficial dose (1 mg/litre) and a deadly one is very large. With current amounts added to drinking water you'd need to drink about 4,000 litres (1,000 gallons) of water within a day or so (fluorine doesn't accumulate in the body) to have a lethal dose.

The picture gets complicated because the harmful effects of some geological materials are seriously overplayed. Asbestos is probably the most feared natural mineral in recent decades, and yet there is little real public understanding of this extraordinary substance. The term describes types of fibrous minerals that occur naturally in rocks formed from ancient ocean crust; in the Mediterranean, asbestos-bearing rocks have been mined in northern Italy, northern Greece, central Turkey, Cyprus and Corsica. The remarkable property of asbestos is the ability of its fibres to resist heat. The fire-retardant properties of asbestos were known as long as 5,000 years ago. It was the ancient Greeks, in awe of the stone that could be woven but not consumed by fire, who gave it its name 'asvestos', meaning inextinguishable. Throughout ancient history, it was renowned as a substance of almost magical properties.

There are six types of asbestos mineral and today they are all officially labelled as harmful. When materials that contain any kind of asbestos are disturbed or damaged, the fibres can separate and become airborne. If significant amounts of the fibres are inhaled they can become trapped in lung tissue, sometimes so severely damaging them that they cannot function. The resulting disease of asbestosis affects those mining or working with asbestos, with the most severe lung damage coming from the five types of asbestos whose fibres disintegrate into sharp needles. Yet 95 per cent of asbestos used in homes, offices and schools is made of chrysotile, the one form of asbestos whose fibres are not needle-like. In such normal environments where asbestos is present, typical measured amounts of asbestos fibres in the air are much too low to be significant (about 0.001 particles per cubic centimetre), even if accumulation over many years is assumed.

■ FOOD FOR THOUGHT

Asbestos is just one of the many rock ingredients that have gone from objects of veneration and wonder in the ancient world to being dangerous and feared in our modern one. Rocks that have benefited societies for centuries can also do harm. The geological environment beneath our feet has provided the building blocks for past civilisations and the raw resources for them to grow and prosper, but it has a downside. As we exploit our planet's natural resources, we are increasingly learning of their far-reaching and sometimes deadly consequences. In future years we will no doubt learn more about the risks posed to general human health by the metals and other precious materials that humans have been working and refining for millennia, but we may also discover more about the benefits they bring.

Our survival as a species is inextricably linked with the rocks on our doorstep. It is from them that we derive the essential nutrients for human existence, and the trace elements that are so crucial for our good health. In some cases these are in the air we breathe, released as gases during volcanic eruptions. The water provides other essential elements, collected as it filters its way down through soil and rocks, but the main goodness of Earth comes from rocks themselves, which weather and rot, turn into soil, and feed plants and start a food chain that has humans at its endpoint. As the next chapter discusses, geological changes in the Mediterranean have had an enormous impact on the food we have to eat.

CHAPTER 5

THE MEDITERRANEAN MENU

A BUSTLING MARKET STACKED with colourful fresh fruit and vegetables and arrays of glistening olives, a glass of wine in a trattoria overhung with vine trellises, a fishing village with the morning catch laid out on the quay, all embody the quintessential flavours of the Mediterranean menu. Though it's no longer exclusive to the Mediterranean – much of it has been exported across Europe and North America, where it's hailed as one of the healthiest diets in the world. The people of Greece, particularly Crete, are thought to have the longest life expectancy of anywhere, followed by southern Italy, Spain and France. Also, crucially, it's a long healthy life – even if Mediterraneans don't actually clock up more years than other people, far fewer die of the chronic diseases that afflict much of the West. How infuriating for those toiling away in more frenetic, less blessed parts of the world to find that a laid-back lifestyle among sea, sand and sun should keep people so healthy!

What is it about this diet that makes it such an all-round tonic? It helps to see the Mediterranean menu as a pyramid: foods most eaten at the bottom, those least eaten at the top. At the base of the pyramid are the foodstuffs that form the bulk of the diet – the cereals and grain products. Whether it's pasta and polenta in Italy, rice in Spain, bulgur wheat in Turkey, couscous in Tunisia or bread pretty much anywhere, cereals and grain provide the carbohydrates that give energy. On the next level up are the fresh vegetables and fruits, along with legumes like chickpeas, lentils and beans, another source of energy-boosting carbohydrates and proteins.

Next comes the magic elixir, intrinsic not only to Mediterranean food but a whole way of life: olive oil. Olives provide Mediterraneans with their main fat intake – and, vitally, it's unsaturated fat. Saturated animal fats such as butter solidify at room temperature, while unsaturated fats like olive oil stay liquid and don't clog us up. That's not all. Olive oil is high in those valuable anti-oxidants that slow down the gradual deterioration of body cells (a deterioration akin to an old car gathering rust), while also being high in the good cholesterol that helps keep cells supple but not the bad cholesterol that blocks blood vessels.

Mediterraneans may consume lots of olive oil, but they don't take too much dairy produce, though the live bacterial cultures of yoghurt seem to be great for health. Towards the top of the pyramid the small intakes of animal foods – fish, poultry and eggs – give proteins and healthy oils and, in the case of fish, more anti-oxidants. Just before the top of the pyramid there's a very thin layer of sweet things – the Mediterranean diet doesn't forbid them, but they're eaten sparingly. At the very pinnacle is a tiny amount of red meat, which when eaten occasionally seems to reduce the risk of various cancers and heart disease and benefits health by lowering cholesterol.

And, of course, with all this pyramid building, there is the need for a regular dose of physical exercise and a glass or two of red wine.

So, we are what we eat, and what we eat in the Mediterranean appears to make us healthy. It seems to be such an integral part of Mediterranean life that it must have been there for ever, enjoyed by the fortunate inhabitants long before swarms of tourists arrived, but in fact most of what we eat in the Mediterranean didn't actually originate here. Even the staple foods – the cereals, the grains, the vines, the fruits and even perhaps the olives – were brought here from outside, and some of the most essential elements of the 'traditional' diet, such as tomatoes, are modern additions. So, what do we really mean by the Mediterranean diet?

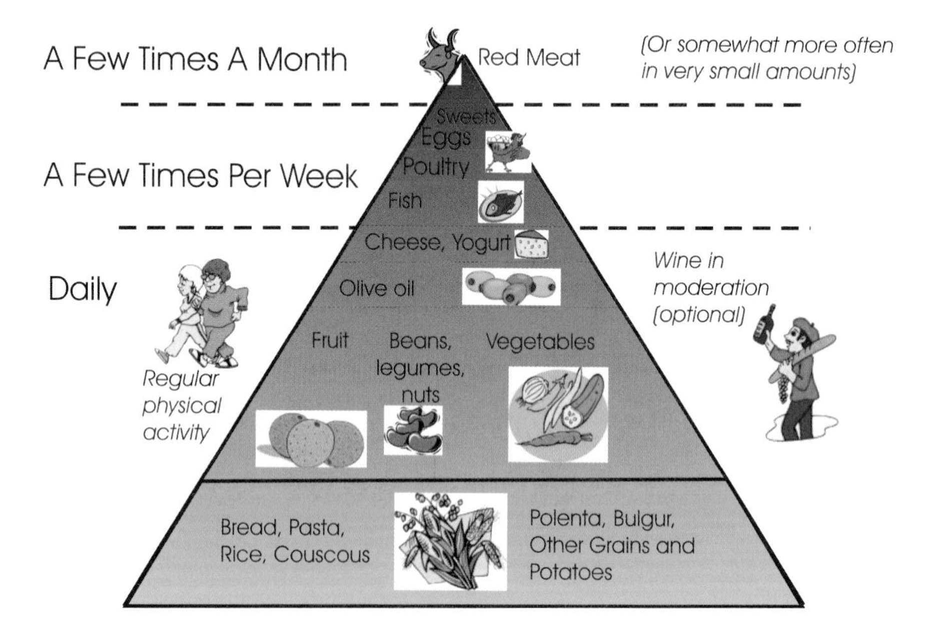

The basic recipe for what we call the Mediterranean diet was invented around twelve thousand years ago in what are today the dusty plains of Syria and Iraq. Here, still in the Stone Age, small groups of nomadic hunter-gatherers began to change their ancient way of life, gradually taking on more and more of the key elements of civilisation as we now think of it: cultivating plants, building permanent settlements and trading in commodities of all kinds. Putting these elements together created a quiet revolution that redefined how we would live, what we would eat, what we looked like, and even how we would die.

What sparked off this great leap forward? If a way of life had suited our ancestors for hundreds of thousands of years, what on Earth would make them change? After all, 99 per cent of human existence had gone before 'Man the hunter' decided to become 'Man the farmer, potter and general jack-of-all-trades'. The trigger was pulled when our planet embarked on a geological makeover . . .

FROM GREENHOUSE TO ICEHOUSE

For those who haven't skied on the snow-capped peaks of Uludag in Turkey, Mount Etna in Sicily or Mount Parnassus in Greece, or for that matter haven't been snowbound on the road to Delphi, the idea of glaciers in the Mediterranean may seem perverse, but in past times snow and ice weren't just winter visitors to the mountains. In fact, throughout most of the two and a half million years of human existence, the world has been in the freezer – and long before we came on the scene, Earth had gone through no fewer than four other Big Chills, roughly 300 million years apart – a pattern that suggests such switches are a fundamental part of the planet's makeup.

Within each of these icehouse periods there has been a constant ebb and flow of cold periods (glacials or ice ages) and warm periods (interglacials). The last two and a half million years have seen around twenty ice ages, each lasting about 100,000 years and separated by more or less ice-free interglacials lasting 10,000 to 20,000 years – and within a single ice age, great continental ice sheets come and go, creeping back and forth across the land.

What makes Earth blow hot and cold? One reason is that the planet wiggles and wobbles. As it spins around the sun, its orbit gradually changes from roughly circular to a definite oval shape, then back again to the approximate circle. This cycle takes about 96,000 years, during which time the planet receives different amounts of solar radiation, and it is this that seems to regulate the start and end of ice ages. Earth is unsteady on its axis, too: one particular wobble happens every 41,000 years, and a second every 22,000 years; these seem to orchestrate the advance and retreat of ice sheets about the globe.

The rhythms of the ice age can also be set by the sun. Even ancient astronomers recognised from observing sunspots that the brightness of the sun was changeable, and modern measurements confirm regular cycles of solar activity over eleven-year and ninety-year periods. Changes in solar radiation may have a profound effect on Earth's climate over decades, even centuries, but over longer periods scientists are less sure. Another way to block out the sun periodically is to throw enormous amounts of dust and debris high into the atmosphere. So, major volcanic eruptions and impacts of comets are popular candidates for causing sudden global cooling; a few big volcanic eruptions do seem to have happened just when the planet was going into a cold period.

A final important regulator of climate is the oceans, and especially the salty currents of the North Atlantic. Today, the Atlantic's equatorial surface waters flow north as the Gulf Stream and meet the cold waters north of Iceland, where the Arctic winds cause them to freeze, forming sea ice. As it freezes, it expels salt. (Arctic explorers walking on the sea ice have always known that it is a great source of fresh water.) The surrounding water becomes extra salty and dense, and this causes the water to sink. A deep current of heavy salt water then flows southwards before meeting a similar but much weaker deep salt current sinking off Antarctica, at which point the two spread out into the Indian and Pacific Oceans. There, in the warmer equatorial waters, the less salty water rises and drives a return surface flow that eventually carries 'fresher' water back into the Atlantic to start the cycle again. It is a long process – it takes a drop of water about 1,500 years to sink down in the North Atlantic and then rise in the Pacific or Indian Ocean – but it is this slow 'conveyor belt' of ocean water that circulates heat around the globe. So, if the sinking of salt in the North Atlantic is shut down, probably by sea ice advancing south, and the ocean stops mixing, the equatorial waters just get warmer and the polar waters just get colder. The result can be the start of an ice age. Alternatively, if the salt current is switched back on again, the return of warm water to the North Atlantic can bring an ice age to an end.

So, through a combination of a wobbling planet, a temperamental sun, the odd erupting volcano and a restless salty sea, our planet has a pattern of flipping from icy to warm conditions. For the last 10,000 years humans have been living through one of those warm periods, though today controversy rages about whether this is the lull before another ice age or whether modern global warming has permanently taken us out of the freezer. What is also not clear is to what extent the successive waves of early human migration on the planet may have their roots in the rhythms of the ice age. What we do know is that the first humans emerged in the rainforests of East Africa two and a half million years ago, just as the planet was starting the latest spell of ice ages. With moisture locked away in the growing ice, a drying climate allowed savanna grasslands to eat into the tropical rainforests, perhaps providing environmental encouragement for our species to 'come out'. After that, the constant switch between cold, dry worlds and warm, wet interludes may have been an important driver for early humans emerging from Africa and spreading across the globe. Their long march would have been made easier, though, by what the oceans were doing.

PULLING THE PLUG

The Corinth Canal, an hour or so's drive west from Athens, may be one of the engineering wonders of the modern world, but to geologists it is a remarkable slice through 350,000 years of changing sea levels. The sedimentary rocks that form the sheer walls of the canal record the restless history of the adjacent Corinth gulf, alternating back and forth from times of high marine conditions during interglacials like today, to dry land or freshwater lake during ice ages. The reason for this bobbing up and down is that to build the colossal ice sheets of Europe and North America, vast quantities of water needed to be withdrawn from the oceans. During ice ages, sea levels across the world fall by around 120 to 130 metres, exposing large expanses of continental shelves as dry land. Then, when the ice ages end, the meltwaters refill the world's oceans and seas up to around their level today. From the southern shores of the Corinth Gulf itself, we now have a record of those rises and falls of sea level that extends back even further than in the canal – a staggering 600,000 years. The landscape here is a flight of steps, each step being a wide ancient beach, deposited at a time when sea level was high and then left stranded by the next sea level fall. As the land has risen, the successive levels of the staircase have been raised high and dry, with ancient shores now exposed to the prying eyes and eager hammers of geologists. The geological traces of enormous sea-level fluctuations may be best preserved in this corner of Greece, but the same dramatic changes would have occurred across the Mediterranean and throughout the world's coastlines.

It isn't known when the earliest of our human ancestors – among them the wonderfully evocative *Homo habilis* and his upstanding successor *Homo erectus* – ventured into the Mediterranean, but many probably came during past ice ages, when sea levels were low. These people are called 'Old' Stone Age (Palaeolithic), after their simple stone technology. Given their rudimentary culture, and the fact that there were so few of them, these people have left little trace of their existence, though human skeletons many hundreds of thousands of years old are found at a handful of sites in south-eastern and northern Spain and in central Italy. This suggests that two important gateways into the Mediterranean may have been through Gibraltar and Sicily. Here, the drop in glacial sea-level exposed islands as stepping stones in the Straits of Gibraltar and Sicily, making the crossings a series of short hops rather than the wide sea passages that they are today.

Even with the shorter distances, these crossings would still have been treacherous for people unskilled in the use of rafts and boats. Instead, most of the earliest Palaeolithic travellers into Europe probably took the long route round the eastern Mediterranean, making use of the broad coastal plains. What is clear is that the bulk of human migration simply bypassed the Mediterranean and Europe, drawn instead to the broader coastal plains of the Near East and south-east Asia. However, it wasn't just that climate and low sea levels had made these plains more attractive – the lands were also on the move.

A GEOLOGICAL YO-YO

Opposite: The Corinth Canal, which cuts the narrow land isthmus connecting the Peloponnese to central Greece, is one of the world's largest continuous natural exposures of geological strata. These strata record how over the last several hundred thousand years the nearby Gulf of Corinth has switched back and forth between freshwater lake and marine seaway as ice sheets have come and gone.

At first it may seem strange that the heartlands of our earliest human ancestors – East Africa, the Near East and southern Asia – were all zones of extreme geological unrest. We tend to think of earthquakes and volcanoes in terms of the devastation they bring, but here we have the earliest centres of human development positively flourishing in these violent places (a point we will return to in Chapter 7). The Mediterranean too is one of the most geologically active places on the planet, and its landscapes are being continually reshaped and renewed. Viewed over the long timescales of human history, even the slow rising of the Mediterranean mountains, or the gradual opening of its seas, have a bearing on how we developed. The wide Gulf of Corinth, the mountains of Calabria, even Europe's highest volcano, Mount Etna, were not around when humans first burst on the scene; all have formed in the last two million years (in the case of Etna, most of it has grown in the last half a million years). Landscapes and humans have evolved hand in hand.

Across much of southern Europe and southern Asia, as well as along stretches of the Near East and North Africa, the ripples from the geological collisions of the past are still being felt. Some parts are still rising up, creating barriers to human movement, while nearby areas are still dropping down, trapping sediment and forming, over time, flat pockets of fertile soil. The faults across which these up-and-down movements take place are routes for water to escape upwards, supplying these land pockets with springs and water holes. The combination of new sediment and fresh water encourages a wide variety of plant and animal species to congregate, and in turn supports a range of hunter-gatherer communities. In the flat, featureless plains it may have been difficult for early human hunters to catch fast-moving grazing animals, but the sediment 'oases' probably helped detain mobile prey. Even the upland barriers provided opportunities. A landscape of rolling hills and valleys could help early people monitor, predict and control animal movements: hunter-gatherers could create bottlenecks where prey could be located and trapped, and places where hunting bands could camp close to, but out of sight of, their prey. It was probably here in the uplands, rather than on the plains, that Palaeolithic man perfected his hunting skills.

LIFE IN THE FRIDGE

With most migrants from Africa bypassing the Mediterranean, modern humans – who left Africa around 100,000 years ago – turned up in the north of the Mediterranean only 40,000 to 30,000 years ago. These peoples, the

Cro-Magnons, found it still in the grip of the ice age. To the north, ice sheets had advanced as far as the southern reaches of Germany and Britain, while sea ice reached the coast of northern Spain. Much of France and southern Europe was covered with tundra, looking much like the Siberian steppes do today. The great oak and pine forests that had blanketed Europe during the warm interglacials were pushed south to occupy isolated pockets in the Mediterranean. Here, scattered woodlands mixed with open steppe-like plains, while the upland interiors were covered in glaciers. Further south the same cold glacial conditions encouraged the Sahara desert to expand north to the coast and south into the savanna grasslands, in turn forcing them southwards to replace the tropical rainforests.

What was life like for these later Palaeolithic Mediterranean peoples? Their stone implements, skeletal remains and fossilised faeces left behind

The snow-covered mountain peaks of the Gran Sasso range in the Italian Apennines west of Rome give an indication of what much of the Mediterranean uplands probably looked like at the heights of the last Ice Age.

reveal much about their way of life, but the most vivid depictions of their world come from their art. The outermost chambers of natural caves provided our ancestors with their first shelters, and their entrances were often adorned with a rich array of images of the hunt – men, weapons and, above all, animals. The Palaeolithic menu is on the wall in the cave paintings at Lascaux in France and Altamira in Spain. From these we see that food came on four legs – the original fast food!

These early peoples, who lived in small groups, hunted down herds of reindeer, bison and mammoth. The grassy plains may have been a dangerous hunting ground for humans, competing as they did with sabre-toothed cats, lions and hyenas, but there wasn't much choice. The cold, dry plains were more or less free of edible plants, so scavenging rather than plant-gathering probably supplemented the unsuccessful hunt. The Mediterranean's small woodland refuges may have offered more opportunities for early vegetarians, and its shores may have provided a small input from seafood, but the Palaeolithic Mediterranean was essentially a meat-only zone.

So there we were – huddling around our cave-mouth campfires, scratching our dream menu on the walls while breaking into hyena skulls to get at the brains. As the future cradle of the western world, the Mediterranean was showing an unpromising start. So what changed? Quite simply, the ice went away and with it went our hunting grounds.

THE BIG THAW

At the end of the last ice age, tourists wouldn't have cruised up the Nile or headed further south to see the Victoria Falls – Lake Victoria was bone dry and its great downstream waters a mere trickle. Neither would they have ferried their way round the Greek Cyclades or sailed the Adriatic between Venice and Dubrovnik, since both areas were dry land. On the other hand, they could have sailed from the Dead Sea to the Sea of Galilee, as a great lake joined the two. Scenic strolls in the Pyrenees, the Abruzzi hills or the Lebanese uplands, however, were out of the question – their mountain peaks were capped in ice and their valleys entrenched by glaciers. It was so chilly that even basking on the beach would have been a tough call. In fact, little of the Mediterranean world that we know today would have existed during the last ice age.

Soon after the height of the ice age 18,000 years ago, however, the ice sheets started to retreat. Between 16,000 and 8,000 years ago they melted away, pouring water back into the world's oceans and raising sea levels. It wasn't a smooth process. As the ice sheets melted, sudden enormous floods of freshwater into the Arctic waters would occasionally shut off the North Atlantic's deep salt current, sending the world spiralling back into a brief cold spell. One of these brought about a return of ice-age conditions 13,000 years ago, but it lasted for only about a thousand years. From around 12,000 years onward, the planet staggered out of the ice age and into a new warm, wet world.

Around the globe, the rising waters were reshaping the glacial geographies, drowning land bridges like the one that had earlier allowed Palaeolithic man to walk from France to Britain. Along the convoluted, island-strewn shores of the Mediterranean, the effect was particularly marked. Low-lying islands, such as those within the Straits of Gibraltar, quickly became submerged and new islands were created. The larger islands like Sicily, Malta, Sardinia, Corsica and Cyprus had always been separated from the mainland by deep sea passages, but now narrow land connections were cut off to places like Corfu, Samos and Cephalonia. The transformation was most striking in the Aegean Sea. Here, from 16,000 years to 6,000 years ago, the Greek Cyclades metamorphosed from a wide, low-lying plain into a pair of landmasses (one centred on Paros and Andiparos, the other on Mykonos-Tinos-Andros), and then into its present scattered rocky archipelago. Throughout the Mediterranean world, the coast was disappearing while many of the island havens that tourists flock to today were being born.

It is the stuff of legends – ancient and modern. With low-lying coasts throughout the world being inundated at this time, it is not surprising that most cultures have oral traditions of a great flood. The flood myths of the Pygmies of equatorial Africa probably stem from the drowning of their Niger delta, while for the Sumerians of Mesopotamia it was the inundation of their coastal marshlands in the Persian Gulf. The Noah's Flood story emerged from the Sumerian epic of Gilgamesh, but the Mediterranean world has a possible candidate for this event – the Black Sea.

Some geologists have argued that at the height of the last glaciation the Black Sea was a freshwater lake, and separated from the Mediterranean Sea by a land barrier on which Istanbul has now sprawled. Then, around 8,000 years ago, the last pulse of glacial meltwater flooded into the world's oceans, forcing a sudden rise in sea level. Around this time, the rising Mediterranean waters are thought to have carved the Bosporus river course through the land barrier and flooded into the Black Sea as a spectacular waterfall, forcing those around its shores to flee. Recent geological studies now dispute this catastrophic cascade, but some still argue that since the roots of most

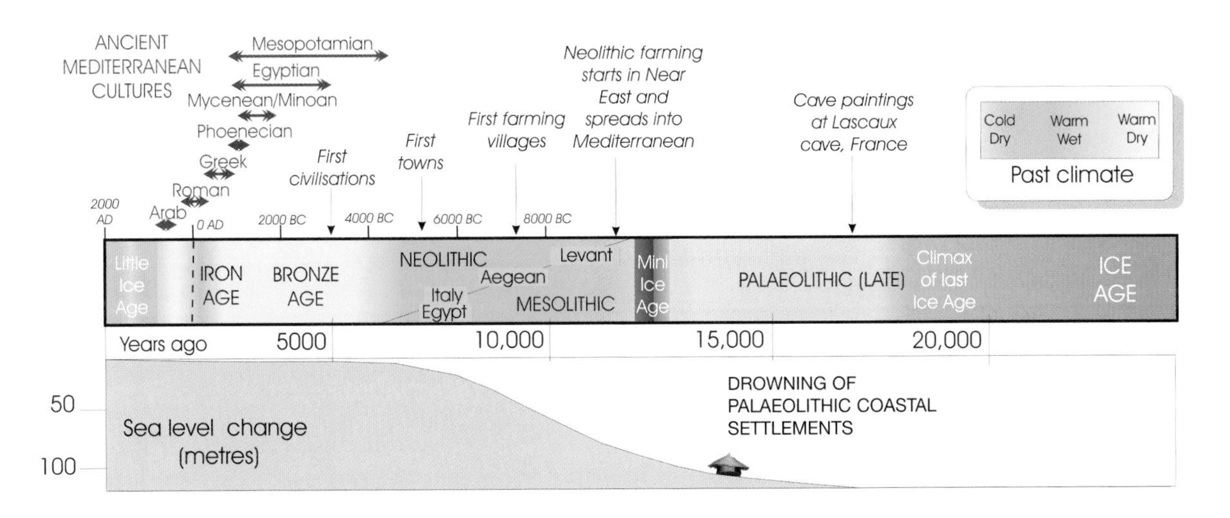

European languages appear to originate in the Black Sea region, the rising Mediterranean waters may have triggered not a great wave of the sea but a wave of modern humans into Europe and the Mediterranean.

Even apart from the possibilities of sudden catastrophic floods, the creeping rise of sea levels had immense implications for human history. When sea levels finally stabilised 6,000 years ago, most of the large hunting grounds of our Palaeolithic ancestors were drowned. The speed at which sea levels rose meant that in many of these coastal plains the shoreline was retreating at around a kilometre per year, forcing people to up-sticks continually and move inland, squeezed into narrower and narrower coastal zones. Much of the Palaeolithic legacy of the Mediterranean now lies sunken beneath its waters, and inland, even more dramatic changes were happening.

Released from their glacial refuges in the south and east, pine and then oak woodlands spread rapidly, replacing the tundra and grasslands of southern Europe. The whole landscape was transforming – rivers swelled enormously, lakes appeared and the summer hills became ice-free. Between 14,000 and 10,000 years ago, most of the humans' large prey animals and some of their predators disappeared. Mammoths, woolly rhinoceros, giant deer and sabre-tooth cats all became extinct, wiped out by more efficient human hunting and by the animals' inability to adapt to changing conditions.

A NEW STONE AGE DIET

For the first time, modern humans were faced with a Europe free of ice sheets. And as the herds of wild game disappeared with the grassy plains, people had to adapt quickly and rethink their menu. Meat still had some role, but now hunting skills were turned to tracking wild herds of bison, sheep and goats as they migrated seasonally from the summer hills to the winter plains. In the open parkland and woodland hills, hunters learned to manage these large mammals. Slowly, hunting became animal husbandry and wild herds became livestock. The hunters and occasional gatherers of the Palaeolithic age were becoming 'Middle' Stone Age (Mesolithic) foragers. The new open forests were full of new edible plants and these gradually shifted the Mediterranean diet away from meat to one of nuts, berries and fungi. These new foods could be collected easily, even by children, using much less energy than that needed to chase meat, and they were almost as nutritious – Mediterranean pine kernels, for instance, have a protein value two-thirds that of lean steak.

Something else highly significant was happening: fire. Palaeolithic Man hadn't so much invented fire as captured it. Natural fires from lightning had been a common phenomenon in the dry ice-age grasslands, but its potential had been only loosely harnessed. Controlled burning had long been used by hunter-gatherers to open up the landscape, but in the Mesolithic, fire was put to work, assisting all aspects of wrenching and coaxing food from the land. Foragers soon recognised that berries, mushrooms and wild grasses flourished best on burned ground, and that a controlled fire exposed acorns and chestnuts. The work that human burning did was much the same as wild fire – it liberated nutrients like calcium, phosphorus, potash and proteins; it restructured the micro-climate of sun and shade, heat and water; it also drove off, if only for a short time, unwanted soil micro-organisms, predators and plants. The Mediterranean flora was designed to burn.

Armed with a flexible foraging lifestyle and with fire in tow, Mesolithic peoples were able to reach more or less all corners of the Mediterranean landscape. Even the ice-free highlands were not able to defy the controlled burning with which they refashioned the environment. The only pockets of native wilderness that resisted were the islands since, with the exception of those lying immediately off shore, they remained beyond the technology of Mesolithic peoples to reach. Free from human disturbance, islands like Malta, Sicily, Cyprus and the Balearics of Majorca and Minorca were arks for native species that had long since disappeared from the mainland. On Malta, elephants came in three sizes: small, medium and large, the last being only the size of a calf today. Among the Mediterranean's other weird and wonderful island menagerie were dwarf hippopotamuses, deer that could not run, flightless swans, and never a predator fiercer than a badger or giant owl.

The origins of these strange *Dr Dolittle* zoos of miniature wildlife lay far to the south, in Africa. They probably arrived as a few lost stragglers wandering north several million years earlier when the Mediterranean was a dried-up wasteland. With the return of the marine waters that refilled the sea, they became stranded, and their island confines saw them grow progressively smaller and smaller. The worlds they inhabited were no lost Gardens of Eden, though. Being free of predators, many of the islands were probably nibbled almost bare by browsing animals.

Within a few thousand years, these island arks would be gone. While tailless Barbary apes still cling to the rock of Gibraltar, today not a single Mediterranean island is left with its native animals. What exterminated them was not Mesolithic foragers, but a new wave of human invaders that would come from the east. For here, a revolution was brewing. And the weapons that they would bring to transform the Mediterranean world were – plants! Or rather, farming.

The Bekka valley in Lebanon, part of the so-called 'Fertile Crescent' of Mesopotamia that extended into Syria and Iraq and which twelve thousand years ago was the rich farming heartland of the Natufian peoples. The Natufian's experiments with growing crops kick-started the Neolithic agricultural revolution that over the next few thousand years spread westwards into Asia and eastwards into the Mediterranean.

A STONE AGE REVOLUTION

A new age was dawning. Across the globe, humans began to experiment with domesticating plants and animals, ushering in the 'New' Stone Age, or Neolithic period. The rise of agriculture, and the onset of the Neolithic, arrived at different times in different places, but the earliest and probably the most important heartland was in the plains of what is today Syria and Iraq. During the ice age, this area had been a refuge for the last natural stands of wild cereals like barley and rye. With the final stutter of the ice age 12,000 years ago and a shift to warm, wet winters and hot, dry summers, the Natufian peoples in what today is Syria began to experiment with the wild

cereals and grasses that flourished in the region, discovering that their seeds could be ground up to make gruel or flour for simple unleavened bread. In what was the world's first GM experiment, these first farmers very slowly began to recognise which of the natural grasses gave the better yields, so they started to grow selected seeds rather than just relying on wild plants. This process of 'unnatural selection' eventually gave us the wheat and barley we depend on today as the staple carbohydrate of all Western diets.

At the time it must have struck some sceptics as ludicrous that certain seeds weren't used to make bread but were instead buried in the ground and allowed to grow; but no doubt they were converted by harvests far larger than before. The growing crop surplus forced the early farmers to set up permanent residences, and feed an increasing population, which in turn allowed more land to be cultivated. The seasonally nomadic hunting and gathering lifestyle was abandoned. The existence of a food surplus meant that some members of the community were freed from cultivation and could instead work in other, specialised, jobs. The need for vessels to store an extra grain harvest brought about a spectacular development in pottery and basket-weaving; and plentiful food could support a labour force that made possible the large-scale working and processing of metals. The demand for clothing resulted in the cultivation of fibre plants, such as linen, cotton and hemp. After satisfying their own needs, a community could begin to trade their surplus for other materials and merchandise they needed. Those first Neolithic farmers had unwittingly triggered a chain reaction that laid the foundations for the complex social and economic organisations that would lead a few thousand years later to the first cities, and then, after another few thousand years, to the first true civilisations.

The fields of grain and grass allowed people gradually to domesticate the animals that browsed on the edge of their settlements. The dog, being carnivorous, didn't have to wait for agriculture and had been domesticated a few thousand years earlier, but generally animal domestication develops alongside plant cultivation. Domesticated sheep appear around 11,000 years ago, closely followed by goats. Pigs were present in agricultural villages about 9,000 years ago, but for the first signs of domesticated cattle we have to wait until around 7,500 years ago.

Animal husbandry offered a reliable food reserve and a wide variety of dairy products – but along with these welcome developments came others much less desirable: disease and social unrest.

Farmers were living in close daily contact with a range of animal species, often under the same roof, and their waste, garbage and granaries were drawing unwelcome scavengers. Suddenly, humans were a rich hunting ground for a startling variety of parasites, and the concentration of more and more people into confined areas quickly brought epidemic and plague. The move to settled agriculture had removed much of the risk around food, relying as it did on a limited number of staple crops and animals, but the Neolithic revolution triggered humanity's first huge wave of disease in its history – the patterns of infectious diseases that we know today can be tracked back to the rise of farming. The reality is that as soon as farming appears, human health deteriorates.

New ways of living, new kinds of settlement patterns and new foods were bringing other unexpected downsides. The human fossil record strongly suggests that as population increases with food production, so the conditions are set for the rise of organised warfare and increased mortality due to violence. Humans have always fought for some reason, but from skeletal remains we see the number of injuries caused by violence increase over time as prehistoric societies began to compete for agriculturally productive lands. As agriculture took hold, life expectancy across much of the Neolithic world dipped from the usual forty years of hunter-gatherers to about thirty.

At first, the idea that Neolithic farmers had a tougher life than their hunting and foraging ancestors may seem perplexing – it's a common assumption that Palaeolithic life was nasty, brutish and short. True, the hunter-gatherers had lived a difficult nomadic lifestyle, moving about the land in their relentless quest for food, but, although they endured a more physically demanding life than the farmers to come, to a large extent they were built for the job, with larger skeletons and thicker bones. Mesolithic foragers would be smaller than their Palaeolithic hunter-gatherer forebears, and the Neolithic farmers would be smaller still. Agriculture was having a profound impact on human evolution.

The most dramatic changes in humans were related to diet. With its emphasis on cereals and pulses, the new diet did offer certain nutritional and health benefits – a moderately high carbohydrate intake, low saturated (mainly animal) fatty acids, and high levels of beneficial fatty acids from vegetables. However, because what we were eating in early Neolithic times was completely different from before, our bodies had to change. For one thing, agricultural societies chewed less than their hunting and foraging ancestors. It wasn't so much a case of what we were eating as how we were preparing it. The new ceramic pots not only stored food but were used widely to cook it – normally tough foods were being reduced to a soft mush. In human fossils, we see a gradual decline in tooth wear (eating is generally getting easier) but, because of the high carbohydrate intake, especially from natural sugars in plants, there is soaring tooth decay. Also, because we need to chew less, our faces and jaws get smaller. The same number of teeth are now crowded into a smaller and smaller space, so dental health problems increase. The roots of modern dentistry are literally to be found in the change in diet 12,000 years ago, but in ancient times toothache didn't mean a trip to the dentist; with nothing to cure the rot even if the diseased tooth was ripped out, it usually meant an agonising death.

Agriculture brought people so many new ways of being killed that it is amazing they survived at all. Even the new diet could be a killer in itself. The carbohydrate-rich foodstuffs were low in protein and in certain minerals and vitamins. A lack of iron is indicated by skeletal evidence – cases of childhood anaemia increased in early farming settlements in the Near East and

Overleaf: The Dead Sea between Jordan and Israel is the world's saltiest body of open water, and as its waters continue to evaporate away faster than they can be replenished with occasional rainfall and the inflow from the Jordan river, it is steadily becoming saltier and dwindling in size. In the recent geological past it has dried up completely, leaving behind great deposits of salt, and if present trends continue a similar fate awaits it in the future.

Evaporating salt pans near the town of Trapani in western Sicily are modern equivalents of salinas established by the Phoenicians almost three thousand years ago. All around the Mediterranean, ancient peoples were transforming natural salt marshes into ponds for evaporating seawater to grow their own salt.

Mediterranean, striking at weaning age when mother's milk was replaced by cereals. Also, the change from a meat-based diet meant that, for all the benefits of a stable, year-round food supply, there was another essential ingredient of which humans weren't getting enough: salt.

■ SALT OF THE EARTH

Salt – sodium chloride – is essential to everyday life. It provides two of the main chemical elements that we can't do without. We need the chlorine for digestion and respiration, but even more important is the sodium. Without it, our bodies would be unable to transport nutrients or oxygen, transmit nerve impulses, or move muscles, including the heart.

Separate the sodium from the chloride and you'd get a metal which left to its own devices would happily burst into flame, and if that didn't hurt you then the chlorine, a deadly poisonous gas, would surely get you. However, combine the two together and eat them and they keep you alive. The average adult human contains three or four shakers' worth, but we are constantly losing it in body functions – urine, blood, sweat, tears and semen all contain salt – and that's the problem. Our bodies can't manufacture the stuff, so we have to add it – although not much of it – to our diets or we'll die a slow painful death.

Today the problem is too much salt in the diet, a product of junk food and its salt-rich additives, but 12,000 years ago, the danger was too little. Previously, our natural intake of salt had been from meat – the large grazing animals having ingested it by licking rocks and soil and drinking from salty waterholes. Those peoples that had followed the reindeer herds north with the retreating ice seemed to have managed to retain their fill of salt. Today, Eskimo peoples like the Sami are the only Europeans left who could live quite happily on a meat-only diet because the reindeer meat contains the right levels of natural salt. For us, however, and our vegetable-loving Mediterranean cousins, the new Neolithic diet meant red meat was relegated to an occasional novelty. Ever since agriculture came, our consumption of meat has drastically dropped and we've had to add salt artificially to our mushy diet – but where does this magic ingredient come from?

One of the oldest arguments in antiquity was about the origin of salt. Was a gigantic bed of salt at the bottom of the sea keeping the ocean salty, or was the Greek philosopher Aristotle right in believing that the ocean waters were fresh and only their surface layers were salty? Aristotle's view that saltwater was formed by the sun's evaporation of freshwater prevailed until the 1660s when measurements finally established that the ocean's deep water was

even more salty than its surface water. But if salt formed in the sea, how then could we explain the great lumps of rock salt that appear on land?

Salt, after all, is just a rock – albeit an unlikely one in that it grows as crystals out of a liquid – and in Neolithic times people mainly dug it straight out of the ground. It got there by rain dissolving minerals out of rocks on land and carrying the dissolved chemical elements to a lake or a sea, where evaporation crystallised them back into solid rock. This process can be seen in action around the disappearing waters of the Dead Sea, where thick salt crusts are building up on a shoreline that is dropping at an alarming rate, about a metre or so each year. In the past the Dead Sea has dried up completely, and what was left behind was a thick bed of salt. Close by the Dead Sea is Mount Sodom, a 6-kilometre-long lump of pure salt caused by the drying up of the great lake. Much of the salty rocks around the Mediterranean come from a vast drying six million years ago, when the enormous seaway turned into a 2,000-metre-deep wasteland of brine lakes and desert. A bed of salt several kilometres thick grew until it was flooded by the inpouring of the Atlantic waters and buried by new sediment. However, salt has a unique ability to flow under pressure and, being less dense than most rocks, it rises up, forcing its way to the surface. Around the Mediterranean, salt mountains have popped up, particularly in places like central Sicily where land upheaval from clashing plates has thrust up the rocks. Today in Sicily working salt mines extract the material, and you can buy six-million-year-old rock salt straight off the supermarket shelf.

Huge quantities of rock salt were extracted in ancient times, but, throughout most of antiquity, the main source of salt wasn't in the rocks, it was in the coastal marshes. Six thousand years ago, with the stabilisation of global sea levels, the world's great deltas began to advance. All around the Mediterranean's shores, great rivers like the Ebro, Rhône, Po and Nile started to rebuild the coastal lowlands, and the extensive coastal flats that developed were perfect pools for evaporating off seawater. What started as natural salt marshes were modified by humans into enormous evaporating ponds and soon, all around the Mediterranean, people were growing their own sea salt.

EXPORTING THE REVOLUTION

Whether salt was dug up from the ground or evaporated off from the sea, the discovery and exploitation of the Mediterranean's limitless reserves of one of the vital elements of human food completed the Neolithic agricultural package. From the Near East, this way of life based on cereals and domesticated animals was able to spread rapidly along the shores of the Mediterranean, its progress eased by the broadly similar climate and geology that stretches east–west from Spain to the Indus. A grass that grew naturally in the Near East was also likely to grow well when taken west. Also, there was often a local equivalent already growing in these new territories which could be manipulated in a comparable way.

With warm, wet conditions flourishing across the Mediterranean, the rural revolution of the Near East had reached the Aegean nine thousand years ago, and by a little later it had spread to the Black Sea coast and west into Asia. By seven thousand years ago it had arrived in Italy and in the Nile delta. The arrival of farming in Egypt seems surprisingly late, given its close proximity to the Near East, but at the same time as the Natufians had been refining their agricultural package, the earliest Egyptians were enjoying a bountiful hunter-gathering life in what today is the Sahara desert. Here, the wet conditions that arrived 12,000 years ago transformed an arid wasteland into a shimmering land of lakes. Monsoon rains on the Ethiopian highlands had refilled Lake Victoria and torrents of water were cascading down the Nile and other rivers.

Today there are visible geological traces of this lost lakeland around the Western Oases. Sticking out of the desert are strangely shaped rock pillars where the ancient lake sediments were carved into bizarre forms by the wind. The bones of crocodile and hippo can still be found, while antelopes, giraffe and fish are depicted in the spectacular rock paintings close to the Dakhla Oasis. Geological studies in this region show that, although domestic animals turn up around seven thousand years ago (but not the camel, which appears only during a dry period three thousand years later), hunting wild animals and gathering wild plants still went on as the first Neolithic farmers were arriving in the Nile delta and spreading south.

As the agricultural revolution spread west, people began to seek out new lands. Building on the design of the simple boats that plied their trade along the Mesopotamian and Nile waterways, and strengthening them, early sailors headed out on the open seas. It was these first seafarers who discovered the lost island arks. At first the islands closest to the mainland were reclaimed but, as seafaring journeys became more ambitious, the more distant, virgin, islands were the destination. Neolithic people landed on Cyprus nine thousand years ago, on Corsica and the Greek Cyclades seven and a half thousand years ago, and then on Crete six and a half thousand years ago. It would take a further thousand years or so for them to reach the more remote islands like the Balearics (Majorca and Minorca), Corsica, Malta and Sardinia. With them they brought the dog – the first carnivore that most islands had seen –

and between them they came, they saw and, by and large, they exterminated. In some places, the removal of the large mammals allowed the recolonisation of trees and plants returning after the ice, but in general what took root were those plants and animals imported with Neolithic agriculture.

Much of what we consider to be the 'typical Mediterranean diet' came from this invasion – virtually none of it is native to the Mediterranean. Of the traditional staple crops in the Mediterranean diet, only a few legumes may be truly native. Some consider the olive to be a homegrown species, but others argue it was native just to northern Israel and north-west Syria and appeared in the Mediterranean in its domesticated form only around six thousand years ago. Cereals and wheat, of course, definitely arrived as part of the Neolithic agricultural package, though they were quickly grown along with native cereals, such as oats which perhaps originated in Crete. The same Neolithic advance brought the vine, a reluctant import since it was used to the wetter northern fringes of the Mediterranean and in the drier south could survive only by constant tending and deep roots. However, as well as spreading westwards, the Neolithic peoples linked eastwards with the emerging agricultural centres of Asia, bringing a counterflow of Asiatic plants into the Near East and Mediterranean, at the vanguard of which were onions, and that other indispensable ingredient of the Mediterranean diet, garlic.

The Neolithic revolution brought more than a few staple crops. It set in place a way of life centred on plant cultivation and animal husbandry that would force Mediterranean peoples to settle and trade and, in time, develop civilisations that would match and possibly eclipse those emerging in Mesopotamia and Egypt. It would be these empires and this international trade that would bring the final elements of the Mediterranean diet. In antiquity the Arabs would bring peaches and figs from the Near East and the date palm from Africa. The Arabs may also have brought pasta, having a version of durum wheat, though it is likely that this evolved in various places – it was certainly around before its 'discovery' attributed to Marco Polo's visit to China in the thirteenth century. In the Middle Ages, apples would be introduced from central Europe, and citrus fruits, rice and sugar cane (along with cotton) would come, via the Arabs, from south-west Asia. With the discovery of the Americas, maize allowed the Italians to have their polenta, and the whole of the Mediterranean took to the least healthy aspect of their diet: tobacco. The tomato, today so strongly associated with Mediterranean cuisine, was a surprisingly late arrival – coming from the Americas with the potato, french beans and sunflowers but not widely grown until the nineteenth century – while the last 150 years have seen the arrival of the aubergine (eggplant) from Africa, banana and kiwi-fruit from south-east Asia and the avocado pear from the Americas.

So, it was these imported foodstuffs that immeasurably enriched the old inadequate Mediterranean menu and made it the delicious as well as health-giving diet of today. Perhaps this way of eating should shed its familiar name and be called, more justifiably, the Global Diet.

However, even as this new way of life took hold in the Mediterranean, it was under threat. The farmers who had spread westwards had done so on the back of a well-watered world, but 6,000 years ago the Mediterranean began to dry up. Water became the single most important limiting factor on food production, and the next chapter in Mediterranean history would be coping with an increasingly capricious climate.

CHAPTER 6

WATER: ELIXIR OF LIFE

LIFE ON EARTH BEGAN IN WATER, and all organisms need a constant supply of it to survive, but in essence the value of water lies not so much in what it is made of as what cargo it carries. Pure water, a barren combination of hydrogen and oxygen, is a tasteless, odourless, colourless substance which on its own is not sufficient to sustain us or most other living creatures. Instead, it is the property of a water molecule to assimilate other chemical elements that makes water the elixir of life. A pinhead-sized drop of water contains a billion billion molecules, each of which efficiently picks up a whole range of rock-forming elements. In taking in other minerals, water provides the wide range of nutrients that living organisms need. The fluid inside all living cells is mostly water – indeed, we ourselves, like most animals, are around 65 per cent water, while some soft-bodied aquatic animals, like jellyfish, are more than 98 per cent – though spiced with proteins, and DNA, sugars, salts, fatty acids and hormones. Similar nourishing cocktails flood the cells and bodily machinery of all living organisms, circulating much-needed sustenance and carrying off waste. Life cannot survive without water, and we humans need to consume at least a litre a day of it for long-term health.

High in the atmosphere, pure water picks up carbon dioxide gas and forms a weak carbonic acid rain which, when it splashes down on Earth, attacks the chemical bonds in minerals and releases the elements. Once it is in the soil, rainwater picks up the hotchpotch of elements vital for life. Natural spring waters may look pure but they typically hold at least twenty-five dissolved elements, such as silica, calcium, potassium, magnesium and iron. These elements, collectively called salts, plus varying minor amounts of gases provide natural waters with the myriad tastes on which bottled water manufacturers make their profits. Too many natural chemicals, however, spoil it. More than a few hundred parts per million of salts and it begins to taste too bitter, and even tiny quantities of organic compounds can make it undrinkable to humans. When it comes to water, human bodies are very picky; the range of water composition on which humans can survive is a narrow one, and that range approximates to that of average river water.

It might seem unfortunate then, that while the Earth's outer layers are dominated by water, only a minuscule fraction occurs as fresh running surface water. Of all the water on the surface of the Earth, just 3 per cent is fresh – that is, contains far lower concentrations of dissolved minerals than sea water, which is always close to 35 grams of salts per kilogram of water. Even then, roughly four-fifths of this fresh water is frozen in polar ice fields or mountain glaciers, while the remaining fifth lies under the surface, as groundwater in the rocks. Less than a hundredth of one per cent is available in freshwater lakes and rivers. Fortunately, that critical tiny amount is constantly being cleansed and replenished, ensuring that the small amount of water available is endlessly recharged, along with its potpourri of dissolved chemicals.

Water is essential to life not only because it is an excellent solvent of nutrients but also because this nourishing fluid can move or flow easily. Organisms have innumerable pathways and transport routes by which water is moved around: blood, lymph and urine are among the bodily fluids that feed and purge internal waterways. For the planet, that vital circulation network is the global water cycle. This continuous recycling of water is driven not by geological forces in the Earth's interior, but by the sun. The sun's heat warms the liquid seawater, which evaporates into invisible water vapour and then dissipates into the atmosphere. Only pure water evaporates, ensuring that any dissolved minerals and substances are left behind, maintaining the saltiness of the oceans. It is estimated that a volume of water equivalent to all the world's oceans passes through the atmosphere every three thousand years. Annually, that is some 380,000 cubic kilometres, enough water to cover the Earth's entire surface to a depth of 1 metre. The oceans are not the only source, since plants lose water vapour during transpiration and animals breathe it out and perspire it away. All this vapour rises up into the higher, cooler parts of the atmosphere, where it condenses as water droplets and eventually falls back to Earth as rain or snow.

Although most water vapour is evaporated from the oceans, the bulk of that moisture is carried by air currents to fall as precipitation over land. The rainwater that drains quickly into rivers and streams will probably be back in the ocean within a few weeks. However, water frozen on the surface of glaciers and ice sheets or seeping deep underground may have a longer detour, spending anything from many centuries to hundreds of thousands of years locked away. It is reckoned that within a week or so of returning to the ocean, most water droplets will rejoin surface waters and be quickly dispatched by evaporation to start the whole cycle again, though some will descend to languish at more sluggish depths.

While the homeopathic notion that water molecules retain a 'memory' of their endless journeys round the planetary block in their atomic structure is unlikely, seawater is certainly older in some parts of the world than in others. The youngest surface waters are in the North Atlantic, but they take a few hundred years to make it to the Mediterranean. Once through the Straits of Gibraltar, it is reckoned that the average water molecule takes a further eighty years to circumnavigate the basin, moving first along the coast of North Africa, then up the Levant shores and then swinging through the

Aegean and Black Seas before heading over to the Tyrrhenian Sea and along the coast of Spain to be spat out, warmer and saltier, at Gibraltar – something to remember the next time you take a gulp of seawater at Marbella!

While water may be endlessly recycled around the planet's surface, that doesn't mean that everywhere gets its fair share. The global water cycle isn't a process that the planet instigated to maintain life, but is a response to the uneven heating of the globe by the sun. The sun's heat is greatest at the equator, where the world heats up, and least at the poles, where it cools down. Ocean currents, such as the operation of the salt conveyor system discussed in the previous chapter, are one way that the planet can move that heat around, but weather is another. The excess heat generated at the equator is circulated by air currents towards the poles. This process, modified by the planet's orbital spin, is what drives our weather.

■ WEATHER FRONTS

Today, reliable summer sunshine is the basis of the Mediterranean's tourist industry, with its cloud-free skies one of the main attractions for the solar-starved visitor. However, the habit of tourists to endure day after day of inexorable heat on an arid beach is only around half a century old. In Mediterranean tradition people have learned to avoid the burden of the heat of the day; they have distrusted the sun, which brings thirst and weariness and heat stroke and destroys the complexion, and have loved shade and the summer night.

From one year to the next, minor variations in the strength of neighbouring weather systems bring slightly more or less rainfall or slightly higher or lower temperatures, sometimes at unexpected times of the season. For the most part these fluctuations are bearable, even welcome. Occasionally they are extreme even for hardened locals – summer heatwaves cause people to drop dead in the streets of Athens or Cairo, or winter deluges sweep all before them across the Midi of southern France or Pyrenean Spain. Sometimes too little water, sometimes too much – these are not just idiosyncrasies of the Mediterranean climate, but its normal limits.

Every summer, the Mediterranean is like an oven. The hot dry sirocco winds waft across southern Europe from the Sahara, flushed out from above. Cold air descending from the upper atmosphere warms and expands as it sinks, to maintain a persistent high-pressure weather system that creates cloud-free skies along the north coast of Africa and keeps out surrounding moist air. The downdraught is a counterweight to a rising current of moist air further south in the tropical lands either side of the equator. At the equator, moisture evaporated out of the warm ocean waters rises up into the atmosphere. As it rises, it condenses, and much of it falls as rain, mainly in intense tropical downpours during the hot season. This low-pressure equatorial weather system, the Inter-Tropical Convergence Zone (ITCZ), is the rainmaker for much of the planet. In summer, the thermal equator – the planet's line of maximum heating – lies in the low latitudes of the northern hemisphere where the large continental masses warm up more than the oceans. Consequently, rain-bearing ITCZ monsoons generated in the Indian ocean nudge west into northern Africa and Arabia, tantalisingly close to the parched Mediterranean. Their path blocked by the highlands of Ethiopia, monsoonal rains drain instead down the Blue Nile to reach the Mediterranean Sea. Over four-fifths of the freshwater reaching the Egypt's Nile valley comes from these Ethiopian rains, but that precipitation is highly seasonal, falling only in June to August when the Indian monsoon arrives.

In the northern hemisphere winter, the Mediterranean weather heads south. The cooler northern landmasses no longer deflect the thermal equator northwards and instead the planet's belt of greatest heat returns south to the geographical equator. As the tropical rains shift south, the pool of stable dry Saharan air follows, settling over the North African interior. Only the southernmost parts of the Mediterranean, and outliers like the Canary Islands, feel the protective warm flush of desert air. The rest is exposed to the mercy of

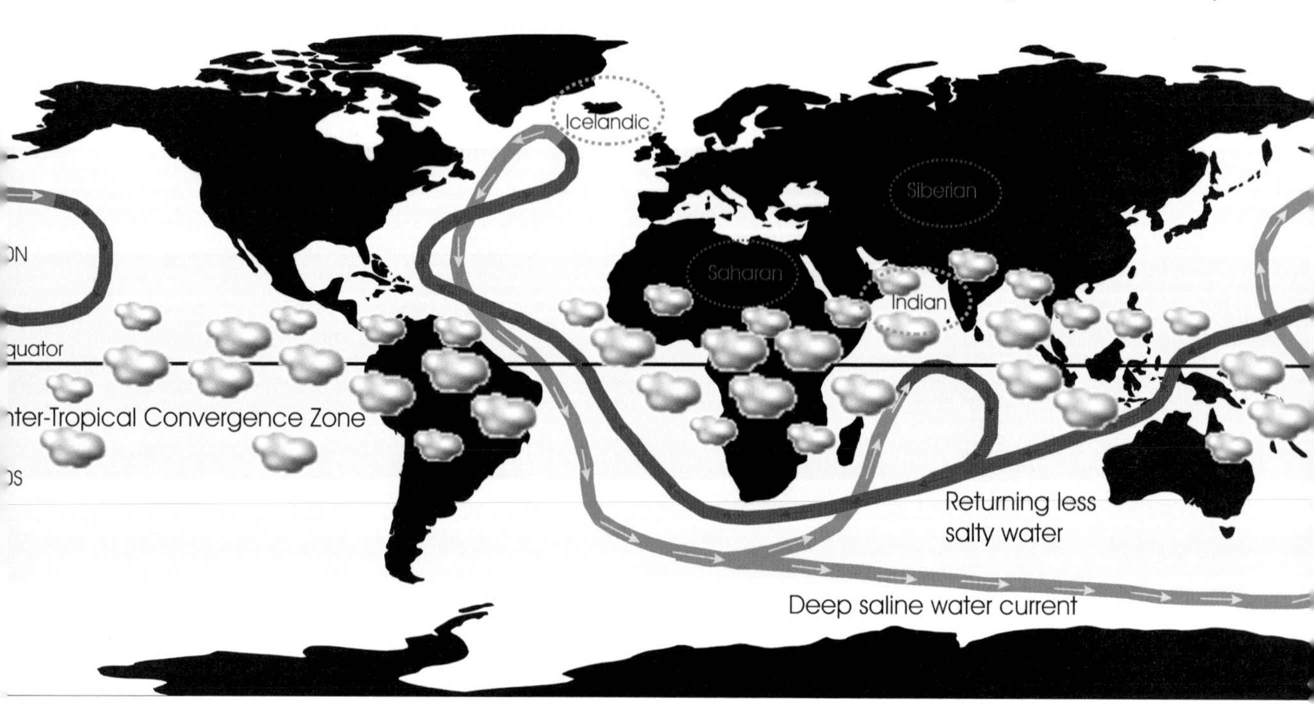

cold northern winds. Most sweep eastwards as a succession of low-pressure weather fronts, or depressions, pick up moisture evaporated off the warm Gulf Stream waters of the North Atlantic ocean and dump it across Europe. Cold, wet Atlantic storms buffet the Mediterranean's northern mountain rim, from the Pyrenees to the Alps, searching for breaches to squeeze through. The Rhone Valley of France is one such chink in the armour, and a moist, cold wind, the mistral, funnels down this, skirting the western Alps and chilling the French Riviera. Further east its sister wind, the bora, breaches the eastern Alps to whip up Adriatic rainstorms. Much of the rainfall in the Mediterranean comes during a brief period of storminess between December and March when the most intense storms sweep eastwards.

The see-saw between warm, cloud-free Saharan weather and cold, stormy Atlantic weather leaves the Mediterranean caught between long, hot, dry summers and brief, mild, moist winters. Other parts of the globe at equivalent latitudes north and south share this seasonal battle between hot and cold, warm and dry, but the so-called 'Mediterraneoids' of California, western Chile, the Cape of Good Hope in South Africa and southern Australia are just small pockets of such climatic conditions in comparison to the great swathe that stretches from southern Portugal to south-west Asia, and from the Alpine peaks to the edge of the Saharan dryland.

Within this broad swathe of 'Mediterranean climate', however, local contrasts flourish. For those who want rain, in general west is best and east is least. The winter rain-bearing westerlies that sweep in from the Atlantic are forced up over mountain ranges, drenching the western windward sides and leaving eastern leeward slopes as rain shadows. Thus Almeria in south-east Spain is Europe's driest place, with an annual average of less than two centimetres of rain, while the coastal ranges of Croatia are the continent's wettest place with a yearly rainfall average of almost fifty centimetres. While it may be that 'the rain in Spain falls mainly in the plain', like everywhere around the Mediterranean, the rains come more in one place than another in one year, then shift their preferences the next and following years. Modern records of rainfall in Spain show no correlation with equivalent patterns in, say, Greece, and there is even precious little similarity in precipitation patterns between next-door neighbours like mainland Greece and the island of Crete. Such local variations in climate have long been crucial in ensuring the survival of Mediterranean peoples in difficult times. Excessively wet and dry years rarely affect large parts of the region at the same time, so areas whose harvests had failed could probably get replacement crops or seed from within a few hundred kilometres. Throughout human history, the Mediterranean's capricious climate has been just one of the many drivers for trade.

It has not always been like this, however. In the past, such extreme conditions appear to have taken hold of the whole Mediterranean.

THE GREAT DRYING

Although the last 10,000 years have not witnessed any of the dramatic changes that signalled the end of the Ice Age and so profoundly changed the lifestyles of our hunter-gatherer ancestors, our planet's climate remains far from stable. The warm, wet world that carried the Neolithic revolution out of the eastern Mediterranean was at a time when the tilt of the planet's orbit allowed the northern hemisphere to receive enhanced amounts of summer heat. High sea-surface temperatures strengthened the Inter-Tropical Convergence Zone, allowing its moisture-laden monsoonal winds to penetrate into the interiors of North Africa and Arabia. Geologists call this period the 'climatic optimum', since across much of the northern hemisphere temperatures were higher than today. The Mediterranean may actually have been slightly cooler due to increased cloud cover and soil moisture, but summers were warm and winters mild. Most importantly, there was plenty of water around, with abundant rains and swollen rivers.

Since the boom times of the climatic optimum, the planet's changing tilt has gradually brought less summer heat. A weakening of the ITCZ gradually drew the much-needed monsoonal rains south and the Mediterranean began to dry. Although the change was slow, imperceptible in human terms, geologists increasingly believe it was accompanied by episodic bursts of deteriorating climate. These short, sharp shocks seem to be related to what was going on in the engine room of the northern hemisphere's weather, the North Atlantic. Evidence from deep-sea sediment cores taken in the North Atlantic suggests that every 1,500 years (give or take five centuries) icy Arctic waters push southwards, cooling the ocean surface and reducing evaporation. The resulting cold, dry conditions trigger a few centuries of mini-ice-age, before normal service is resumed in the North Atlantic and moisture-laden westerlies again deliver their much-needed rains to southern Europe. Quite what keeps this quiet climatic pacemaker ticking away now that the great ice sheets have gone is uncertain. However, the suspects are the same as those most implicated in putting the planet as a whole in the freezer: periodic slowdowns of the North Atlantic salt-conveyor system or periods of reduced heat from the sun. What is clearer is that these cold flushes in the North Atlantic produced major societal change in the Mediterranean.

The Mediterranean began to dry up 6,000 years ago, or at least that is when we start to see the effects. Up until about 4000 BC, summer monsoons had maintained rainfall levels in Mesopotamia 25–30 per cent higher than the meagre amounts that today make this a desolate wilderness of salt-encrusted desert. Ninety-five per cent of the discharge of the great Euphrates river came from rains delivered to the Anatolian mountains by North Atlantic westerlies (as they still do today), but as the climate changed the monsoon rains drifted ever more south and the northern rainstorms that fed the rivers weakened and were increasingly too late to water summer crops. Frequent droughts descended on the lands of the Near East and, as today, the response of Mesopotamian farmers was to abandon parched fields and move elsewhere. Some went back to the nomadic existence of their distant ancestors, but most began to congregate along the main waterways.

In Egypt, it was a similar story as intensifying droughts settled over the Nile valley. In the south, Neolithic herders grazing cattle on what had before been rain-fed grasslands gradually retreated into the highlands of East Africa, where they thrive today as the Masai cattle people. Most farmers, however, moved north in large numbers to where there was the only source of reliable water – the Nile valley. However, with the loss of the monsoon rains from Egypt, the Nile river too had become unpredictable. As the great drying wiped out vegetation across southern Egypt, when the monsoon rains fell in the Ethiopian Highlands the parched land was unable to absorb the surface waters, which instead charged down the Blue Nile. Something like 80 per cent of the water reaching Aswan in southern Egypt came from this burst of Ethiopian precipitation. Despite the dangers of the autumn torrents, more and more people crowded along the river. Farming settlements along its banks collaborated to build and maintain drainage ditches that diverted floodwaters on to their fields, and to put far more of the swampy land under cultivation. Villages coalesced into walled mud-brick towns, which then began to trade and compete with each other, developing into small autonomous communities.

■ THE BIRTH OF CIVILISATION (AND PLUMBING)

Around 3200 BC, two centuries of extremely cool, dry conditions gripped the eastern Mediterranean and in doing so set the scene for the emergence of western civilisation. The severity of the conditions forced drought-hardened farming communities across the region to capitulate; villagers flocked to the growing urban centres along rivers like the Euphrates and Tigris and, to the south, the Jordan and, of course, the Nile. It was during this short climatic crisis that the world's first cities and states emerged in the region of

Sumeria in southern Mesopotamia. Such states found that with their centralised governments and populations of several thousand people, they had the key elements necessary to tend the vast waterways and maintain the labyrinths of narrow canals, ditches and sluices that now carried the lifeblood of a river-fed world. In Egypt too in 3200 BC, a single ruler, Menes, unified the disparate territories and began to oversee the construction of basins to control the floodwater, digging canals and irrigation ditches to reclaim marshy land; he even diverted the course of the Nile to build his capital city of Memphis on a reclaimed flood plain. It was the 1st Dynasty of Egyptian civilisation.

Since the times of the earliest civilisations, the wise management of precious water resources have been critical to survival. Today that practice continues, here building water control schemes in the Wadi Mujid in Jordan, a canyon prone to catastrophic flash floods.

Irrigation had got the fledgling states of Mesopotamia and Egypt through the extreme droughts of 3000 BC, but the pattern was set for the future. For the next few thousand years, water would be the crucial raw material of the countryside. In landscapes far less well watered than before, the ability to get enough water to irrigate crops adequately became the single most important limiting factor on food production. A new breed of Mediterranean farmers began to develop strategies to cope with an increasingly unpredictable climate. Over centuries and millennia to come, the main field of technological development would not be in military prowess, but the less exciting arena of plumbing. The management and control of water supplies became a prime area for human ingenuity. One of the first machines in history was a simple lifting device, the shadoof, for raising water out of a well or river. It consisted of a balancing system that lifted water in a container which was then emptied out into an irrigation canal. Around 2000 BC, one person could draw about 2,700 litres of water in a day.

Some of the greatest engineering projects of the day were not the enormous monuments being constructed, but the dams that were turning whole river valleys into reservoirs. The Orontes Valley in Syria was closed off by a dam two kilometres long. In Egypt, the prediction of the date and height of the annual Nile floods was so crucial that it was the job of the high priests. A series of flood gauges, called Nilometers, were placed along the entire course of the river – the earliest scientific measuring instruments. Strictly speaking, the Nile fields weren't so much irrigated as inundated, a typical rise of river level of eight metres effectively covering all the farmland. The Egyptians simply managed the floods, directing the rising waters through flood gates and sluices into rectangular reservoirs, which varied in size from 4,000 to 16,000 hectares, and allowed the water to soak into the soil. After fifteen to twenty days of soaking, the dikes were breached and the water flowed first into lower reservoirs and then back into the Nile. The annual flushing out of the fields meant that there was no build-up of harmful salts in the soil, leaving only fresh nutrients. It was this annual water cycle that gave the Egyptians their calendar of three 'seasons', each of four months: inundation (autumn), growing (winter) and harvest (summer).

While the Nile floods were largely regular, the waters of the Tigris and Euphrates were far more erratic. The Euphrates is a low-gradient waterway meandering for almost 2,500 kilometres across northern Syria, with major settlements concentrated along its naturally raised banks, or levees, which can be a kilometre or so wide. As well as the Anatolian rains, it is fed by the seasonal input of wadi flashfloods and is thus prone to flooding and sudden shifts in its course. The Tigris is shorter and steeper and, fed by large upstream tributaries, its upper stretches are turbulent and dangerous. Here water could be safely lifted from the river only by means of the shadoof. The downstream floods of both rivers were hugely dangerous, and their waters tended to arrive at the beginning of summer, too early to water the land sufficiently.

Consequently, the Mesopotamians were forced to embark on a vast project – that of continuously irrigating their lands. Between the lower course of the Tigris and most of the Euphrates, a series of canals was built that criss-crossed the Mesopotamian plain; the largest was 120 metres wide and 300 kilometres long. This web of waterways, the ancient levees of which can be detected in satellite images from space, were designed to carry river water through a branching network of smaller canals and ditches to each individual plot of land. They were amazing feats of engineering, ensuring that the level of the main canal was always higher than the ground level, and using gravity to regulate the flow – too fast and it would destroy the embankments, too slow and the water would pool and stagnate. Care had to be taken to avoid a build-up of salts on the surface of the fields, a problem even more acute in the coastal marshlands where seawater flooding was frequent. Continual flushing of the cultivated lands by irrigation schemes was so vital to successive Mesopotamian empires that special officials were appointed to ensure canals were clear of rushes and waterweeds, waterways were dredged of silt, and banks were strengthened against floods. Later, in 1760 BC when Hammurabi, ruler of the first dynasty of the Babylonians, established the world's first set of common laws, there would be a special provision to punish those who neglected the canals.

The obsession with irrigation would ensure that for much of Bronze Age times, the great civilisations of Mesopotamia and Egypt managed to keep their heads, literally, above water. Elsewhere in the Mediterranean, societies depending on rain-fed agriculture were struggling. A decrease in average rainfall of about 30 per cent is sufficient to cause major upheavals in agricultural and hydrological systems.

■ CIVILISATION IN CRISIS

Between 2200 and 1900 BC, the transition between the Early Bronze Age and the Middle Bronze Age, a particularly severe global cooling and drying event appears to have coincided with one of the most tumultuous events in the history of the Mesopotamian Fertile Crescent. In the face of severe drought, most of the kingdoms in the region collapsed, and major towns and cities were abandoned in the dry farming plains of what is today north-eastern Syria and northern Iraq. Written records refer to this period as the time of 'seventeen kings living in tents', implying that regional political control had broken down, and people were reverting once more to their ancient nomadic traditions.

Nomadic peoples like the Amorites displaced by the drought descended on the rich irrigated heartland of southern Mesopotamia, where the cities of the Sumer and Akkadian empires doubled in population, and the number of settlements soared. Much of the kingdom succumbed to tribes emerging out of the desert wastes in the north-west and north-east. In 2030 BC a 180-kilometre-long wall, the 'Repeller of the Amorites', was built from the Euphrates to the Tigris across the northern edge of their territory to keep the nomadic peoples out. Eventually the last king of Sumer and Akkadia was dispossessed in a period characterised by famine and rising prices. A flourishing civilisation was replaced by a more primitive one.

In Egypt the times were equally momentous. The period from 2200 to 2000 BC saw the collapse of the Old Kingdom dynasties as the land descended into political instability and economic disaster. Individual pharaohs' reigns for this period followed in quick succession – allegedly seventy kings in seventy days at one point – and the economic and political resources of the state were severely reduced. The Nile flows were much reduced and, with the failure of the annual floods, there were crippling food shortages and famines. State building projects collapsed: only two pyramid construction projects were planned – but abandoned.

The effects of these three centuries of drought were felt across the region. Populations suddenly swell along the Indus river valley of northern India, suggesting the influx of people into the better-watered irrigated lands. In the central plains of Anatolia, settlements virtually disappear, and in mainland Greece and Crete there is evidence of a population crash leading to large parts of the countryside becoming deserted. In Greece, as in many parts of the eastern Mediterranean, vegetation died from lack of rain and consequently, when flashfloods came, the fertile soils were stripped off and dumped in the valley bottoms. At Knossos, the ancient political centre of Crete, community cisterns and other water storage structures appear, centuries before the great palaces of the Minoan heyday. From India to the eastern Mediterranean, it seems that three centuries of drought toppled towns and cities and left the world's first empires crumbling.

Perhaps the most resonant account of this great climatic crisis is a story that many of us are familiar with. The biblical account of the destruction of Sodom and Gomorrah is an archetypal lesson in morality that reverberates down the halls of time. According to the account, Abraham and Lot were among a band of semi-nomadic people migrating to Palestine from Mesopotamia, settling in the hilly region near the Jordan valley. Lot chose to settle his cattle herds close to the city of Sodom on the well-watered plain of Jordan, comparable to the intensively cultivated 'garden' of the Nile valley. However, after a breach of morals by the inhabitants of Sodom, God rained fire and brimstone down on the cities of the plain, destroying Sodom and nearby Gomorrah with all their inhabitants; only Lot and his wife were permitted to survive (though when Mrs Lot looked back at the destruction she was turned into a pillar of salt).

Behind the biblical moral drapery is an account of the transformation of the agricultural richness of the Jordan valley into a wasteland. The popular modern explanation of the destructive event is a great earthquake, of which there is certainly no shortage along the Jordan valley, but turning fertile plain into a land of barren desolation is difficult to attribute to purely seismic effects and is more consistent with a dramatic change in climate. The timing of the event seems to place it around the middle of the Bronze Age climatic crisis, since biblical scholars generally consider Abraham to have lived some time between the twenty-second and eighteenth century BC. Archaeologists studying the ancient Palestine of this period found that urban society collapsed, with walled towns being replaced by unwalled villages, cave occupations and campsites. Some areas became entirely devoid of settlement, while others survived only in places along the Jordan river where the stream flow did not disappear. Overall, it is estimated that the population of the Jordan valley was reduced to a tenth of what it was in earlier times. In the Dead Sea and further south in the Red Sea, the period coincides with an increased deposition of salt, implying increased evaporation and decreased input of freshwater. All this evidence suggests the severe drying-up of the whole of the Middle East region over a period of several centuries. The rapid collapse of a civilised society must have left deep impressions on those who survived the devastating climatic transformation. Perhaps a great earthquake during this turbulent time did provide the coup de grâce to an already collapsing society, but it is also likely that with the destruction of Sodom, Gomorrah and the other cities of the plains a myth was born that would leave its mark on western theology and culture for millennia to come.

It took the return of the rains and a few hundred years to allow the ancient world to recover from the 2200 BC climate catastrophe, and for urban life and political stability to return to great centres of Mesopotamia and Egypt, but while more and more people crowded into ever larger settlements dependent on a dwindling water supply, the ancient world had shown itself to be more vulnerable to climate than ever before. In the past, changes in climate had marked the great leaps forward. Ten thousand years ago a warming Mediterranean world had helped our foraging ancestors turn into farmers, and five thousand years later the great drying encouraged villagers to become city dwellers, but now, a drying climate was threatening empires with extinction.

For some scientists, this extinction occurred around 1200 BC when the Bronze Age abruptly ended. The finale of the Bronze Age world is a spectacular cultural implosion as, across the eastern Mediterranean, empires and city states crumbled within the space of fifty years. As we'll see in chapter 7, throughout the Aegean, Anatolia, Cyprus and the Levant, most of the famous cities, flourishing trade centres and rich palaces met a violent end, among them Mycenae and Troy. Smaller communities were not destroyed but many were abandoned, and across the region there was a series of major crop failures that triggered large-scale migrations of refugees. Egypt escaped the catastrophe (and sent grain to the famine-stricken areas) yet in Libya nomads had to move into Egypt's settled lands in search of water. In Mesopotamia there was no significant damage.

The scientific debates as to what caused this regional catastrophe seem to be almost as tumultuous as the events they attempt to explain. Theories range from invasions by a nomadic warring people and changes in the nature of warfare, to natural causes such as earthquake storms and cometary impacts, but one long-standing theory argues it was down to the invisible forces of climate change.

Clear evidence for a dramatic change in environmental conditions at this time is found in the trees. Their growth rings provide an annual indication of environmental conditions that can be counted back through living trees and then through ancient timbers for many thousands of years. Across north-west Europe, these tree records show that around 1160 BC there were practically no rings, indicating a marked deterioration in climate. In south-west

Anatolia, by contrast, trees dated to the same period were positively exploding with growth, with abnormally large growth rings. The tree rings seem to support an argument which was put forward over forty years ago that a northward shift of the dry desert winds of the Sahara desiccated the lands of mainland Greece, Crete and Anatolia. Climatologists have subsequently shown from modern meteorological patterns that Greece is actually on the boundary between regions of deficit and excess moisture, meaning that dramatic rainfall differences occur from one region to the next. In particular, when there is a strong winter trough of low-pressure conditions over the western Mediterranean, unusually dry conditions settle over the Aegean; crucially, south-west Anatolia also becomes wetter while Libya becomes drier, as does northern Syria. In short, all the areas affected by drought correspond to the main areas of destruction.

To many geologists and archaeologists, it seems likely that a few years of extreme drought in the Aegean may have been the trigger for a wave of famine, disease and social unrest that left the cities of the affected region in flames.

GREEK WATERWORKS

With the end of the Bronze Age, a colourful and thriving chapter in Mediterranean history closed. The region descended into what is known as its Dark Ages, when the advanced civilisations of the past were replaced by more primitive societies. The scanty evidence we have about climate during this time suggests there may have been a return to cooler conditions, with the northern lands witnessing advancing mountain glaciers and retreating forests. Across Europe, 1200 BC to 700 BC was a period of cool summers, mild but wet winters and great storminess. The centuries between 750 BC and 500 BC seem to have been particularly wet and stormy. In the Mediterranean, the rains tended to appear in late autumn and early winter, and in threatening abundance. Thus, while what had preoccupied farmers in Mesopotamia and Egypt was saving water, in Dark Age Greece they were more concerned about draining away its dangerous excesses.

Some of the earliest Greek waterworks are the impressive cisterns that Minoan and Mycenean settlements had constructed in the 'dry times' almost a thousand years before. These domestic rooftop cisterns, capable of capturing about half of the rainwater falling on buildings, would become the model for the municipal reservoir, like the rock-hewn tanks that the Nabatean peoples of Petra in Jordan would make five hundred years later to harvest run-off from rocky cliffs rather than rooftops. Such structures were designed largely to preserve meagre water supplies, but later Greek societies would need more advanced water control systems to cope when the climate got wet. Intriguingly, many of these water management projects of Dark-Age Greece may be commemorated in its ancient folklore, in particular in the famous Labours of Herakles (Hercules to the Romans).

Herakles, it seems, may have gained his fame not for his exploits as an early superhero but as a hydraulic engineer! Although some of his Twelve Labours took him to the ends of the known world, most are set in his Greek homeland of the northern Peloponnese. From all across Greece there are stories of Herakles's apparent waterworks, manipulating the paths of rivers, turning marshland into fertile plains and building drainage channels to prevent valleys becoming lakes. In one labour, Herakles settled a border dispute between neighbouring territories in central Greece by building channels and

Vista across the Argive Plain from the Bronze Age citadel of Mycenae in the Greek Peloponnese. On the left horizon is the Gulf of Argos and beyond, the Aegean Sea, though in Myenean times the shoreline was far closer to the city. In the distant right is the site of the ancient Lake Lerna, the marshy ground famed as the home of the Lernean Hydra that Heracles vanquished as one of his Labours.

Lake Stymphalos in the mountainous interior of the Greek Peloponnese where Heracles cast out the predatory Stymphalian birds in what might be regarded as glorified swamp reclamation project. Many of Heracles' famous labours can be interpreted as water engineering projects, carried out at a time when the climate of Greece was cold and wet.

embankments along the river Acheloos that separated the two but whose path was continually shifting because of excessive silting. In another he cleaned the Augean Stables by diverting a river to wash them free of 'debris'. Other feats may commemorate heroic attempts at reclaiming swampland. The defeat of the multi-headed Lernean Hydra seems to have been a tale of how multiple sources of a spring which fed a notoriously marshy area were 'tamed' by Herakles, blocking their entrances and thereby drying out the place. His defeat of the man-eating Stymphalian Birds, with their unfortunate habit of dropping poisonous guano that blighted crops, is a major deed of water control, since with their removal the swampy plain of modern Lake Stymphalos apparently became cultivatable.

The importance of adequate water control in the apparently wet times of ancient Greece cannot be overestimated. As outlined in Chapter 4, during this period iron technology was spreading rapidly across the Aegean world,

giving farmers new tools to farm with. Across Greece, farming settlements were putting more and more of the land under cultivation, and in the wet climate much of this would have been waterlogged particularly when the seasonal rains came. With the emergence of iron farming tools during this Dark-Age period, agriculture would eventually feed the rise to the various independent Greek city-states, and the golden age of Ancient Greece.

In antiquity, some of the most prosperous parts of Greece would be the region of Arcadia in the northern Peloponnese, which grew vines, wheat and barley and supported the large cities of Orchomenos, Mantinea and Tegea; today all three are grassed-over ruins. Here, water management would be used in a very different way – as an act of war. Following Herakles's vengeful precedent of turning Orchomenos into a lake by flooding the plain, the practice of flooding an opponent's territory in order to gain advantage was not uncommon during the Greek conflicts. The most famous incident, the Spartans' defeat of Mantinea in 418 BC, is a classic example of how a simple appreciation of geology can be a real asset.

Today, the grassy ruins of the former Greek city of Mantinea, ten kilometres north of modern Tripolis, belie the scene of the largest land battle of the Peloponnesian War (and, it turns out, a favourite ancient skirmish of modern war-game enthusiasts). In 418 BC, a force of Spartans several thousand strong, led by their king Agis, advanced on the plain towards Mantinea as the inhabitants took up positions on the hill above their city. Concerned with the locals' strategic advantage, Agis decided to withdraw his forces and use geology to win the day. The king got his men to divert the waters of the river Sarandapotamos ('forty streams') so that they flowed down on to the landlocked plain on which Mantinea stood. With the autumn rains imminent, the Mantineans realised that the rising river water, combined with any rainfall, would soon flood their fields. They came down on to the plain to engage the Spartans and were defeated. They had lost out to a secret weapon – Agis's knowledge of 'karst'.

Named after the high limestone terrain of Kars plateau in Croatia where it was first discovered, karst is landscape where there is a virtual absence of surface drainage. Instead of flowing along streams and rivers, water seeps through the rock to be channelled along a subterranean drainage network. The key element of karst is limestone which, being permeable, lets water percolate down through it. As the water seeps down from the soil, carbon dioxide picked up from organic matter makes it aggressively acidic, allowing it to dissolve its way through the calcium carbonate rock. Cracks become enlarged into fissures and eventually into gaping vertical pipes called sinkholes (also dolines or swallowholes). The Greeks gave these features the name

'katavothra' and considered these natural holes into which surface streams disappeared to be sacred places. It was these katavothras that played such an important drainage role in early Greek times, being created or unblocked in the case of Herakles reclaiming swampland at Stymphalos, or being deliberately plugged in the case of his flooding of Orchomenos. It was Agis's realisation that the sinkholes that drained the Mantinean plain would be unable to cope with the excess water coming from the river that led to his idea to flood his opponents out. Half a century later Iphikrates, the renowned fourth-century BC Athenian general, tried the same trick at Stymphalos, hoping to force the locals to surrender by blocking the main katavothra but being apparently thwarted by divine intervention.

Sinkholes are just one part of a vast karst landscape that covers the Arcadia region of the northern Peloponnese. Most of the sinkholes occur at the edge of fertile plains that sit in flat depressions surrounded by limestone hills. These natural basins, generally many kilometres across, are called poljes and they have formed by the dissolution and collapse of large areas of limestone rock that once lay above. The largest polje in the region is the flat treeless plain that Tripolis itself sits in, drained by five major katavothra. Many of these sinkholes lead downwards into cavernous passageways, and then into enormous cave systems. These subterranean worlds often display magnificent natural architectures and sculptures. As the water seeping through cracks drips down from cave roofs, the change in pressure and temperature conditions compels the carbon dioxide to come out of the solution and leave behind the calcium carbonate, which crystallises. A slow drip rate encourages stalactites to form on the roof, and a faster drip rate favours the build-up of stalagmites on the cave floor. Elsewhere, the walls are decorated with calcite in weird and wonderful calcium carbonate formations – curtains of rock hanging from the roof, irregular branches of cave coral, and cascading sheets of flowstone. However, it is not only water that flows here – one sinkhole receives the sewage waters of Tripolis, discharging them out into the nearby Gulf of Argos. A cave to be avoided!

The largely hidden karst landscapes of Arcadia are by no means unique in the Mediterranean. The widespread nature of limestone across the region means that most countries have their fair share of magnificent caves and related karst landforms. However, it was the ancient Greeks who, with extensive tracts of karstic rocks under their feet, first learned to appreciate its importance as a natural plumbing system. Early Greek philosophers like Thales and Aristotle formulated concepts on how the subterranean waters moved through the landscape, and centuries later these would be tested by practical experiments to find out exactly where the waters went. At

Opposite: The vast interior plain of Feneos in the central Peloponnese is a natural amphitheatre formed by the dissolution and collapse of limestone rocks in the centre. Numerous similar land-locked basin pockmark the mountainous Peloponnese interior, each one drained by sinkholes that carry surface water underground only to emerge at springs many tens to a few hundreds of kilometres away. Throughout antiquity when the sinkholes were blocked, or when their capacity was overcome by winter rains, the fertile farming plains became waterlogged swamps or lakes.

Previous page: Beneath the streets of modern Istanbul lie hundreds of gloomy cisterns left from the days of Constantinople, the largest of which is the Yerebatan Saray cistern. Built by the emperor Constantine in the 4th century and enlarged by Justinian in the sixth century this subterranean reservoir had the capacity for 80,000 cubic metres (over 21 million US gallons) of water. It was constructed during the late Roman period when many parts of the Mediterranean world were experiencing a drier climate.

Stymphalos, the second-century AD traveller Pausanius noted that the Greeks used to toss pine cones into these so they could work out where the water flowed to. It was apparent even in antiquity that the waters sinking into the katavothras of the mountain plains emerged as freshwater springs in distant valleys or even at the sea. As we now know from the Tripolis sewage, most of the poljes of Arcadia drain eastwards below the mountains to discharge at the springs of Lerna and Kiveri along the Gulf of Argos, though some in the north drain all the way to Olympia on the west coast.

At first many Greek cities were situated close to freshwater springs to ensure a plentiful supply of water. The ancient Greeks became expert at finding hidden water, learning to spot the physical clues such as the plants that prefer to grow their roots in concentrations of water. In the Mediterranean area these plants are often the fig tree, the rosemary bush, the ubiquitous bramble and the more obvious moss. There were geological clues too. Whenever impermeable clay lies above or below a layer of permeable limestone or sandstone, water collects in the seam and seeps out as a spring.

Not satisfied with where these natural springs emerged, ancient Greek engineers soon began to fine-tune them, artificially enlarging passageways in the rock and setting up networks of wells, cisterns and pipelines for more effective water storage and distribution. Often these springs were linked directly to fountain-houses with long basins to catch and hold the water, and to sanctuary buildings where water was required for rituals. Water outlets became the hubs of the ancient Greek cities. Later, from the sixth century BC, impressive engineering works like tunnels and aqueducts indicate a new determination to bring water from a distance. No longer need the city be tied to the spring.

By the end of the Greek era, cities were supplied by various different water sources: a community could draw on its wells and cisterns, on fountains within the city supplied by springs from inside or outside, and on larger public reservoirs on the outskirts. Not only had they learned how to find water and to transport it long distances but much importance was placed on using it wisely. Cisterns in courtyards collected water funnelled from roofs and stored winter rainwater that would last a household the six-month summer season. Cistern water was used for washing bodies, dishes and clothes, while spring water was preferred for drinking. The waste water and storm run-off were collected in sewers draining out through slits in the ramparts or channels under the streets leading through the gates of the city. This water was used for irrigation, thereby replenishing the water table and ensuring continual running of springs.

The Greeks ingeniously used their understanding of the natural water

cycle in karst terrains to design how water could be circulated through their cities. Their feats of hydraulic engineering assured most cities with a regular supply of water even in dry years, and soon they were spreading their water management skills around the ancient world. Colonies sprang up in Sicily, western Turkey and the Black Sea, generally in limestone terrain where appreciation of water supply was best developed. That they generally chose well is evident from the success and long life of many of these colonies; the most striking is perhaps Byzantium-Constantinople-Istanbul, which the Greeks began 2,800 years ago.

With over three hundred marble columns, the spectacular Yerebatan Saray cistern, or Sunken Palace, displayed all the water management skills of the ancient Greeks and Romans.

BOOM AND BUST

Karst landscapes are ideally designed for the capricious Mediterranean climate, acting as long-term water storage systems that soak up the irregular rainfall like a giant sponge and then release it slowly but surely over thousands of years. Outside these limestone territories, however, most regions still relied on rivers to supply their water needs, and these were far more susceptible to the short-term vagaries of a changeable climate.

After the cold but stormy times of early Greece, the ancient world began to warm up and dry out. Based on the various plant species described by ancient writers, the climate of Classical Period Greece (fifth and fourth centuries BC) seems to have been virtually identical to the present day. Over the next thousand years, conditions would steadily improve until by the late Roman time (fifth century AD) they were warmer than today. As temperatures in the Classical world slowly climbed, Alpine glaciers melted, retreating from 300 BC to AD 400 and allowing the Romans to work gold mines in the Austrian Alps that are today buried under ice. It was in this new warm world that the armies of Alexander the Great swept into Asia and Imperial Rome then expanded to take over much of that territory. Accounts from the first century BC to the second century AD point to moderate warmth with a much wetter climate regime. A weather diary written in Alexandria (Egypt) in AD 120 reports the occurrence of rain in eleven months of

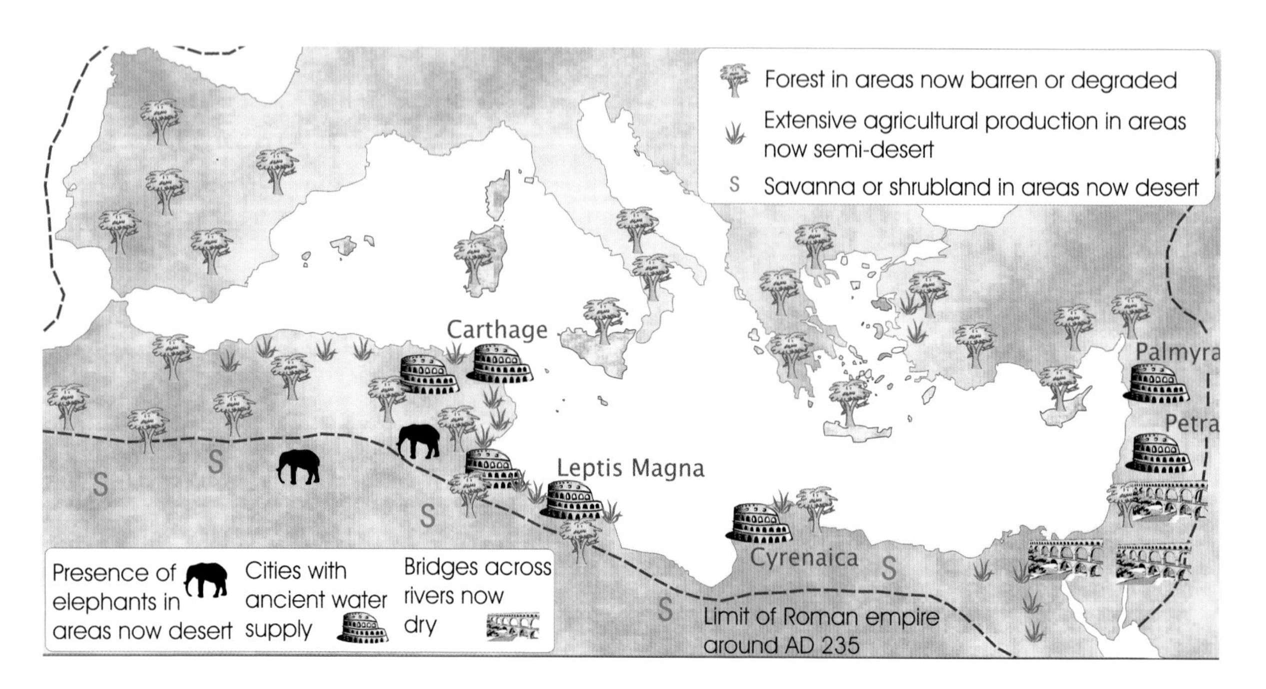

The Claudian aqueduct near Rome was one of nine artificial waterways that served the imperial city, together delivering more than 300 gallons a day water from pristine lakes and springs in the Alban hills a hundred kilometres or so away. Around the Mediterranean, the Roman empire enjoyed a far wetter climate than the region endures today.

the year and frequent thunderstorms throughout the summer. Baths and aqueducts were supplied from sources that do not exist today and bridges over rivers from Italy to Arabia that no longer flow are all indications of a wet world very different from that of today.

Just how different is indicated by the account of Pliny the Elder of elephants roaming the forested southern slopes of the Atlas Mountains, emerging in winter to graze over rich pastures. Today these areas are completely desert, but back then much of the area was scrubland or so-called Mediterranean forest – evergreen oak mixed with dwarf palms, cypress and pine. Broad-leaf deciduous oak forests covered much of Spain, Italy, France and the Balkans. The lush vegetation of the southern Mediterranean at this time appears to have had a critical role to play in maintaining the moist conditions. The presence of grasslands and forests across northern Africa ensured more of the sun's heat was absorbed rather than reflected, and temperatures rose. As ground temperatures rose and evaporation levels increased, air pressure fell over the land, thereby drawing in moist air from the Mediterranean Sea. The lands of the southern Mediterranean – Iberia, the Atlas region, the Sahel and the Nile Valley of Egypt and Sudan – were all moister. It seems likely that wet monsoonal winds of the Inter-Tropical Convergence Zone were once again drawn into Mediterranean Africa.

The warm, wet world offered a real boom of agricultural activity. In the last century BC, cultivation of the olive and the vine had spread north into parts of northern Italy where in previous centuries winters were too cold. Similarly, in the uplands of south-western Turkey, the discovery of olive presses and olive pollen from the Greek–Roman period indicates that olive orchards were growing far above the modern altitudinal limit for olive cultivation, implying temperatures 2–3C degrees Celcius higher than today. Although North Africa was warm and moist, there was only limited agricultural activity; the Carthaginians, for example, still got most of their grain from Sardinia and Sicily. Rome itself was at first largely self-sufficient, banning the import of oil and wine produced outside Italy to protect the agricultural interests of its own farmers, but with an Imperial world bursting with a Roman population two or three times larger than a few centuries before, and with large and long-serving armies requiring huge supplies of grain and wine, the Emperor Augustus (27 BC–AD 14) soon lifted the ban.

In the first century AD, with most of the Mediterranean under their control, the Romans converted vast areas of North African forest or shrubland into olive orchards or grain plantations. During the reign of Julius Caesar, the Roman city of Leptis Magna, whose magnificent ruins now lie in semi-desert wastes of Libya, supported a tax of three million pounds of olive oil. Northern Morocco, Algeria, Tunisia, Libya and Egypt soon became the most productive and thriving breadbaskets for several centuries. North Africa became the granary of Rome, supporting six hundred cities (compared to Gaul, Roman France, which claimed a mere sixty cities), most of which had thermal baths, proof of the ready availability of freshwater.

To satisfy the demands of agriculture and settlement, more and more of the Mediterranean forests were burned and cleared. In fact this had been going on since the arrival of Neolithic farmers many millennia before. Initially, the farmers of the Early Bronze Age had probably preferred to cultivate more stable soils on the gentle slopes at the centre of the plain, but by the Late Bronze Age people had extended cultivation to the less stable slopes on the foothills, steeper slopes that were more susceptible to erosion. Land clearances accelerated during the Dark Ages of Greece, and by Classical times the effects were clear to see; the philosopher Plato lamented how the water had carried off 'the soft, thick layers of earth' around Athens and 'all that remains is the bare carcass'.

In Roman times, this exploitation of the land in many areas went into overdrive. The face of Italy was changed as forests were cleared and drainage schemes undertaken. In the Roman colonies loss of land accelerated in the second century BC with centuriation – the practice of dividing land up into small farms of similar size by rectangular grids of roads and ditches – which lasted for the next four centuries. Roman agricultural writers cautioned the detailed way that various crops needed to be alternated to allow the soil to recover, and advocated the use of small farms and personal ownership of holdings to avoid overburdening the land, but from the first century AD this 'environmentally concerned' approach faltered with the extra deforestation resulting from the allocation of land grants to returning war veterans, and with the rise of latifundia – large villa estates owned by an elite few. The owner of a small plot of land applying the basic rules of crop rotation in order to allow the natural recovery of soil nutrients was gradually replaced by the large estates engaged in monoculture.

But perhaps it wasn't all the Romans' fault. There is some evidence that the end of the Roman period was marked by a long and continuous drift back to drier conditions. Rome's river Tiber flooded only twice in the period AD 174–489, compared to over twenty times in the preceding three centuries of wetness. The Romans, it seems, were having to battle the same difficulty that bedevilled previous great empires – a Mediterranean that was drying up. But it was a vicious circle: the more natural vegetation and forest cleared, the more reflection of the sun's heat and the less evaporation available to feed rainfall, so the climate dried even more and cropland failed. Nevertheless, by the third century AD, vast areas like the plain of the river Po had been stripped of their forest and scrubland and converted to farming.

The clearing of huge areas of forest outside the fertile river valleys exacerbated a problem that was endemic to the Mediterranean – soil erosion. The trees had performed the vital function of binding the soil together, so when rainfall was no longer broken by the forest, water hit exposed ground and took the fertile topsoil with it, washing it into the rivers. The rivers simply weren't powerful enough to carry all this extra soil and eroded rock out to sea so it was dumped at the rivermouths. Many of the Mediterranean valleys began to clog up.

Migrant labourers toil to remove vegetation covering the once magnificent colonnaded boulevard that led down to the former harbour at Ephesus in western Turkey. In ancient times it would be the inability of the Ephesians to keep the harbour free of river sediment that gradually forced the abandonment of what was once the greatest trading centre in Roman Turkey.

DELTA BLUES

The Mediterranean coastline had been steadily advancing for 6,000 years, ever since the rise in global sea levels abated, though periods of climate downturn seem to have given occasional bursts of impetus. The advances were greatest in coasts whose hinterlands were being pushed up by on-going tectonic movements, and the clogged-up river valleys and muddy deltas became the fertile cultivated fields of the Roman empire. However, sediment being washed off the cleared hills in increasing amounts was ensuring that these lands were still on the move, and claiming the coast, entombing harbours and ports in the process.

Some of the greatest cities of antiquity fell before the advancing might of

the rivers. Ancient Troy, the great harbour city famously besieged by the Greeks (probably around 1250 BC) now lies landlocked, ten kilometres from the coast of the Dardanelles Straits in north-west Turkey. Sediment carried by the ancient Scamander river slowly infilled the estuary, creating the river plains and delta swampland on which the Trojan war was fought. Today its ruins and the battlefield are entombed beneath at least ten metres of river sediment. It was a similar story further south, where the great Roman ports of Ephesus, Priene and Miletus succumbed to the inexorable advance of the same fertile muddy plains that had made them rich in the first place. Ephesus – in its second-century-AD heyday the 'metropolis of Asia Minor', with a population of 350,000 – shifted position in an attempt to survive, Emperor Augustus relocating the Roman city seaward of the original Greek settle-

Where has the sea gone? The view from the Roman amphitheatre at Ephesus looking down the line of the harbour boulevard to the marshy plain that now buries the former port area. Today, the continuing deposition of river sediment had pushed the modern coastline eleven kilometres away, leaving the city's land-locked ruins as one of Turkey's greatest archaeological treasures.

ment, but to no avail. By the late sixth century AD, despite continual dredging of the harbour, Ephesus was effectively cut off, hobbling on for only a few centuries more via a winding canal that maintained the narrowest of lifelines to the sea; today its mothballed ruins are one of Turkey's greatest archaeological treasures while the modern coast lies on the western horizon, eleven kilometres away.

Opposite: Once arguably the greatest of the Seven Wonders of the Ancient World, a solitary column is all that is left of the Temple of Artemision at Ephesus. Its waterlogged ruins occupy one of the many stagnant pools of water that surround the ancient city and which in antiquity were breeding grounds for malarial mosquitoes. The spread of malaria around the late Roman world no doubt contributed to the collapse of the once great empire.

The silting up wasn't just affecting Rome's overseas colonies. At home, port cities on the Italian mainland were becoming landlocked towns. For instance, Luni (near Pisa), which had flourished by supplying the local Carrara marble to Rome, had its quarries abandoned and its harbour silted up by the sixth century AD. More significantly, the port of Ostia at the mouth of the Tiber, which handled Rome's grain imports, was abandoned, also because of silting. All around the Roman world, silting left a graveyard of harbour cities and with them died the trade and grain supplies that they delivered.

The loss of revenue and resources wasn't the only problem that the rivers gave the Romans. As bustling harbours were relegated to quiet backwaters they became stagnant water pools that were breeding grounds for mosquitoes that brought malaria. The disease, which left people weak, apathetic and short-lived, had been around for centuries, affecting Greece in the fourth century BC and striking Rome in AD 79. Native to African rainforests, it had travelled down the Nile to the Mediterranean, then spread east to the Mesopotamian Fertile Crescent and north to Greece; Greek traders and colonists then carried it to Italy (from where the Roman armies took it as far north as England and Denmark). By the fifth and sixth century AD much more of the Mediterranean's fertile coastal lowlands had become malarial swamps. In AD 431 it was reported that at Ephesus large numbers of the main Christian leaders had become ill with malaria but, racked by disease, the depopulated city limped on, many of its early Christian churches abandoned. Many of the Roman ports around the Mediterranean became rife with malaria, among them Ostia.

Whether it was mud or mosquitoes that claimed the lives of numerous ports in the Roman world, with each harbour city lost, the prosperity of the empire faltered. As the western half of the Roman world crumbled in the face of northern barbarian incursions, its capital was moved from Rome to Ravenna in AD 402. For some historians the choice of this obscure town on the edge of the Po delta was a strange one, surrounded as it was by marshland. However, unlike the pestilent Tiber marshes left behind in Rome, the coastal lowlands of Ravenna were malaria-free; regular flooding of the marshes by the sea meant they were too salty for mosquito breeding grounds (though by the Middle Ages these swamps too were malarial).

Then, in AD 476, when Ravenna was conquered by barbarian invaders, the western Roman empire too sank into the marshes.

■ DAM, DAM, DAM

Despite the dramatic advances in the Mediterranean coastline during historical times, they are as nothing in comparison to the way that the coastal geography of the region has been reshaped in recent centuries. The sixteenth and seventeenth centuries are periods of particularly strong delta growth. The mouth of Rome's Tiber river pushed seaward, fed by major deluges; there are said to have been ten floods in the second half of the fifteenth century there and thirteen in the seventeenth, compared to no more than six in all the preceding centuries AD. On the Po river, a stretch of delta front tens of kilometres long advanced between eight and eleven kilometres during the seventeenth century. Similar accelerations, though not of this magnitude, are recorded for this period at the mouths of many Italian rivers.

The river mouths of Spain and Greece have been on the move too, though here the eighteenth and nineteenth centuries seem to mark the greatest advances. At Thermopylae in central Greece, for instance, the site where in 480 BC the three hundred Spartans famously held back the Persian army's advance at the narrow coastal pass, the site of the battle is now on coastal plain five kilometres broad and buried under twenty metres of mud; most of the delta growth here occurred in the nineteenth century.

As ever, the causes of this modern phase of delta advance are varied and complex, but many geologists lay the blame at the door of the Little Ice Age. This event was the planet's final (so far!) cold fling, beginning around AD 1315 with devastating storms and ending in the 1740s with a final burst of floods and hard winters. In the Mediterranean region the intervening centuries were typically characterised by cool summers and cold, snowy winters, during which there were repeated episodes of extreme weather events and river flooding, and regular disastrous harvests, famines and plagues. Sediment washed from ruined fields unattended by a devastated populace is commonly cited as the reason for the advance of the deltas around this time. Much of the later advance is attributed to a variety of causes, mostly clearance of forests to fuel the Industrial Revolution and increased management of the great river courses, as embankments built to retain their floodwaters forced ever more sediment directly to the river mouths.

Whatever the cause of the great delta advances, erosion and floods have given the modern Mediterranean the most fertile farmland, the best sites for hotels and airports, and some of the best wildlife habitats in Europe – but in the last few decades there are signs that the sea is fighting back. Many deltas and coasts are no longer advancing, they are retreating, and with them are going one of the region's most important natural attractions – its beaches.

Unlike the Romans, who suffered from too much material being dumped along their shores, many modern coasts are being blighted by water washing away their beaches. Beaches are maintained by a very delicate balance between rivers bringing sand and gravel down to the coast and the sea moving that sediment along – but the development of tourist resorts, with their long ribbons of hotel complexes and concrete promenades, have cut many river mouths off from their supply of sand from the interior, and those rivers that still do flow to the sea are often starved of the power they need to bring down significant quantities of sand. And the culprit is – dams.

Since the 1950s the Mediterranean has embarked on a frenzy of dam building, mainly to provide local communities with electricity and to store much-needed water against the vagaries of climate, and the dams are having a devastating effect on the coast, trapping much of the sand and silt and reducing the power of rivers to carry it down to the coast. The finer sediment – silt and mud – still gets through, but that's not much good for making sandy beaches.

Many coasts are also losing sand because developers have obstructed 'longshore drift', the important process that distributes sediment along the coast, creating and nourishing beaches. The delicate balance of the coast is upset. Many modern harbour structures, such as marinas, and concrete sea fronts and defences are bad news for the beach. They impede the nearshore currents that replenish this beach with sand, resulting in the natural beach actually disappearing completely. The result is that many of the beaches along the southern Spanish coast now have sand that has been dredged up offshore and dumped; these beaches are completely man-made.

Many of the new artificial beaches are not robust enough to stand up to storms that regularly strike them. A huge one hit the coast near Marbella in March 2003, and thousands of tonnes of sand were washed away by powerful waves. To replenish one of the most popular tourist spots, 50,000 cubic metres of new sand were needed to replace what had been lost – that's about 3,000 lorry loads. Overall, local authorities along the coast had to find 60,000 lorry loads of replacement sand. Most was dredged up from offshore. It's an ongoing battle with the sea which the coastal authorities are now faced with, year in, year out.

The difficulties faced in southern Spain and along vanishing parts of the Mediterranean shoreline are just the latest skirmishes in a never-ending war

with water which has been going on for six thousand years, and which every so often has claimed countless lives in droughts, floods and plagues. It is a conflict that shows no signs of ending.

The drying Mediterranean world now finds itself part of a planet whose atmosphere is being warmed up by human activity, and some of the threats that water has in store for future generations will be explored in Chapter 8, but while climate, and through it water, will continue gradually and often imperceptibly to shape the course of human history, Earth has up its geological sleeve the potential for huge, irresistible upheavals that can strike in an instant.

Opposite: A tourist beach at Marbella on Spain's famous Costa del Sol, but many of these sandy stretches are completely man-made, being artificially constructed and fed by dredging of sands from offshore. The culprit for the erosion of this coast's natural beaches is the many tourist developments that are disrupting the geological processes that deliver and distribute sediment from the mountains to the sea.

CHAPTER 7
ROCKING THE CRADLE

THE CLIMATE HAD been dry and the rains had been eagerly anticipated, but not this. This was weird – unlike any rain they had seen before. Droplets of light stone were coming from the sky, a grainy mist enveloping the streets and dousing the broken city in a thin blanket of hot gravel. It coated the stone wrecking ball that had earlier been demolishing crippled mud-brick multi-storey tenements. It dusted the workers who were shoring up the toppled stonework and broken beams. The clouds that dropped it had billowed suddenly out of the azure blue sea in the centre of the crescent isle, the ground rumbling as the plume emerged. It was the even greater rumbles in the days before that had broken the city, the collapsing buildings sending people fleeing to their boats in panic. A few had returned to start clearing up, and now this . . . As the rocky rain continued to fall, the repair teams dropped tools and headed off.

The idyllic setting of Nea Kameni in the centre of the island of Santorini (ancient Thera) is one of the most impressive geological sights in the Aegean. And yet this peaceful tourist hot spot was once the scene of one of the most catastrophic volcanic eruptions in human history.

Whether or not they survived the carnage that was to follow is unclear. All that was evident was that no bodies were found when the city of Akrotiri was exhumed by archaeologists more than three and a half thousand years

Opposite: A geological wake-up call. Between the lower reddish-brown soil of the Bronze Age inhabitants of Thera (Santorini) island and the upper whitish layer of pumice fragments are a few centimetres of yellowish pumice. The yellow band is the tell-tale sign that the first ash and pumice erupted out the volcano lay on the ground for several weeks before the main catastrophic volcanic debris rained down, time enough it seems to allow the islanders to flee.

later. The streets and houses that had bustled with 30,000 people were found empty, entombed in over ten metres of ash and debris that perfectly preserved the island's main settlement – a remarkable time capsule of life in Bronze Age Greece. That island was Thera – now Santorini, the volcanic centrepiece to Greece's idyllic Cycladic island archipelago – and the event that froze the city for eternity was a catastrophic volcanic eruption, probably about 1628 BC, though the date is contested. Many archaeologists prefer a later date of around 1500 BC. It was the largest eruption that civilisation had seen, and would be a rude revelation to the Bronze Age world that droughts and famines were not the only ills that Earth had the capacity to deliver. This would be the first of many times that the cradle of the western world would be rocked by geological violence.

It is unlikely that the Cycladic inhabitants of Thera knew they were living on a volcanic timebomb. True, they undoubtedly bathed in the warm foul-smelling offshore springs and grew wealthy cultivating the crops that thrived on mineral-rich volcanic soils, and they probably mused on the unique doughnut-like shape of their island, but this particular volcano had been silent for the previous five thousand years. There were no legends or traditions of a restless beast beneath their feet. So, it was more than a surprise when damaging earthquakes rocked the island and then the first showers of pumice fell. Launched from the 'hole' in the centre of the island, that initial plume of pumice – frothy lava full of trapped bubbles of gas – was the signal that the eruption was starting.

The slightly yellowish colour of the pumice that fell in the earliest stage of the eruption, forming a layer a centimetre or so deep, indicates that it lay on the ground exposed to the air to become oxidised. The level of oxidation suggests there were only a few weeks before it was covered, but it was time enough to allow the last inhabitants of Akrotiri, and presumably the other Bronze Age settlements on the island, to make their escape.

When the main fireworks began the effects were immediate and cataclysmic. A spectacular crater, or caldera, was formed in the northern part of the island and almost twenty cubic kilometres of volcanic debris were blasted out, the force probably triggering great sea waves – tsunamis – that ravaged the coastal towns around the eastern Mediterranean and left the southern Aegean sea filled with floating pumice. Enormous eruption clouds were launched high up into the stratosphere, dumping up to 50 metres of ash on Thera, and up to a metre of ash on Rhodes. In fact Theran ash can also be traced over Crete, as well as in parts of the Black Sea, the Anatolian interior and in the Nile valley, Syria and Israel. Earthquakes are thought to have struck towns as far away as Minoan Crete, where many of the palaces witnessed fire and destruction. Thera itself became a barren volcanic moonscape and, with some of its Cycladid neighbours, would remain abandoned until the Phoenicians set up colonies in 1000 BC, but, despite the trauma on the Cyclades, it was on distant Crete that the most significant effects would be felt.

In the middle of the second millennium BC, Bronze Age Crete was the centre of the flourishing Minoan culture. Its civilisation, with its Linear A script that still defies decipherment, had acquired a sophistication that rivalled its trading partner, ancient Egypt. The Minoans also traded heavily with the islanders of the Cyclades to the north. Then, about 1450 BC, two centuries after the Thera eruption, the Minoan civilisation on Crete collapsed when almost all settlements on the island burned down, and the Myceneans from mainland Greece took over. It is an intriguing question as

to whether the Thera eruption could have triggered the eventual collapse. Certainly, although the immediate aftermath of the catastrophe was traumatic, the Minoans at first seemed to survive, but some archaeologists and geologists have tried to explain the later demise of Crete in terms of a delayed response to the volcanic fallout. It is possible that the volcanic dust thrown high into the atmosphere may have caused a major deterioration in climate, producing a so-called volcanic winter. Tree rings across Europe show a sudden slow-down in growth at 1628 BC, consistent with a sharp cooling and drying of climate. Archaeologists have speculated that the huge amounts of ash and pumice floating in the sea could have led to deoxygenation of the surface waters, causing runaway algal blooms, and contaminating the sea life and passing toxins up the marine food chain into the stomachs of the fish-loving Minoans. Certainly the high fluorine content of Theran ash may well have been poisonous. Daily fluorine intakes of 1 part per million are recommended today, but these lavas have about 150 parts per million. If drinking waters are polluted with more than ten times the daily limit, serious health effects can result, not least of which are osteosclerosis and other disabling skeletal diseases, and eventually death. If water supplies had become toxic, then the loss of farmers and livestock could certainly have led to severe agricultural collapse.

Opposite: Volcanic deposits from the eruption of Thera volcano around 1628 BC cover much of the modern island of Santorini. The sharp horizontal line at the foot of the cliff represents the Minoan ancient land surface just prior to the catastrophe, whilst the fifty metres of white pumice and ash above are the debris that rained down over the course of a few days of cataclysmic eruption. The immediate effects of the enormous blast were felt throughout the Aegean and beyond, but its lingering after-effects may have led to the downfall of Minoan civilisation in nearby Crete two centuries later.

Disasters are a complex business, however; they have subtle, creeping consequences. Some archaeologists argue that the volcanic eruption was simply the catalyst for social and political change, low harvests triggering a chain reaction of economic hardship, then growing civil unrest. In ancient societies there was a tendency to blame disasters on the sins or weaknesses of the leaders; in Egypt pharaohs were executed after natural catastrophes. In Crete the disillusionment with the ruling elite is detected in the gradual abandonment of minor settlements and the reduction of the number of working palaces to one: Knossos. Ritual sites are given up in favour of more primitive cave sanctuaries, burials become less significant, implying an emphasis on living rather than on death, and 'crisis cults' emerge, some with cannibalistic tendencies. It all suggests a weakening of the social and political glue that holds a society together. With little or no trade with the devastated Cyclades which had supplied raw and luxury materials, the Minoan economy probably went into terminal decline. It all ended in a wave of anarchy that saw the great palaces set alight and destroyed the Minoan powerbase. Then, with the Minoan navy in tatters, the Myceneans became the strongest sea power in the region, opening the way for their gradual takeover of the island and its absorption into the wider Greek world. One culture's catastrophe may be another one's opportunity: with the fall of the Minoans, the Mycenean culture, glorified by Homer and the epics of the ancient Greek myths, rose to prominence.

FOG WARNING

The eruption of Thera is an illustration of how, despite the popular depictions of rivers of lava and glowing clouds of fire, it is the slow insidious effects of volcanoes that are the greatest threat to societies and civilisations. Stromboli is a case in point – the most captivating of the volcanic Aeolian Islands in southern Italy, largely because it has a permanently active volcano. Its fiery rage is best seen in darkness, and many visitors to the island take an evening hike to the summit to watch its crater launch periodic showers of incandescent lava into the night sky. Geologists refer to this relentless fireworks display as 'strombolian' activity. It offers sailors out on the Tyrrhenian Sea a natural beacon in the night, hence the island's affectionate name of the

'Lighthouse of the Mediterranean'.

So, all harmless fun it would seem – but, although the volcano last really erupted in anger in 1930, when it left six dead and prompted an exodus of most of the 3,000 inhabitants, it has a hidden darker side. For alongside the firework displays the volcano belches out huge clouds of sulphurous fumes. Normal everyday – strombolian – activity releases an average of six hundred tons of sulphur dioxide, which exceeds that of all the industrial pollution sources in Greater London and equals about one-third of emissions from the UK coal-burning power stations. Added to the enormous quantities of hydrogen chloride that the volcano emits, Stromboli becomes the largest single source of atmospheric pollution in Europe. Not only that, but Stromboli's vents have been ejecting plumes of sulphur, chlorine and fluorine gas with almost clockwork precision, meaning that this pollution has been churned out at these levels throughout the duration of human history. In a world that is increasingly concerned with the health effects of air pollution, not to mention the effects of acid rain on ecosystems, volcanoes provide a surprising additional concern. For the most part, Stromboli's pollution effects are likely to be restricted to the local area, wafting dense pockets of foul-smelling air across the adjacent lands of Sicily and southern Italy and beyond to create local smogs and other unpleasant weather. However, when major eruptions suddenly release vast quantities of sulphurous gases into the atmosphere, the effects can be lethal on an unimaginable scale.

In terms of casualties of volcanic eruptions, the most dangerous activities in the Mediterranean are the dry fogs created when plumes of fine volcanic aerosols are ejected high into the upper atmosphere. These lingering veils of sulphuric acid vapour partially block out sunshine and disrupt weather systems. In the northern hemisphere, mean air temperatures tend to drop a few tenths of a degree Celsius in the five years after major eruptions – 'tend', because the disrupted atmospheric circulation patterns triggered by large eruptions can induce complex and unexpected meteorological effects. For instance, the year after the 1815 eruption of Tambora in Indonesia is widely referred to as the 'year without a summer', a consequence of the cool, wet conditions in western Europe that caused failed harvests, although Scandinavian countries reported a famously dry summer and bumper crops. The cooling that often follows eruptions can result in widespread crop failure in the chilly, rainy summers that come after the volcanic fogs. The resulting severe famine can then lead to disease in both animals and humans due to lowered physical resistance. This starts a vicious circle as the loss of farmers has a knock-on detrimental impact on harvests in following years, leading to widespread famine, then epidemics and possibly pandemics of plague.

Of the seven most intense volcanic dry fogs to have struck the Mediterranean over the last two thousand years, all but one triggered famine and disease pandemics that affected the region within one to five years after the eruptions. An eruption of Mount Etna in Sicily in 44 BC, for instance, caused a dry fog that lasted ten months, causing poor harvests, famine and a severe pestilence across most of Italy and in Egypt. However, many of the Mediterranean dry fogs are products of major eruptions on the other side of the world. Volcanoes as far afield as Tambora, Laki in Iceland and El Chichon in Mexico have erupted sulphur-rich plumes that have triggered catastrophic famines and pandemics in the Mediterranean. The result of the El Chichon eruption of AD 1258 was plagues raging across Syria, southern Turkey and Iraq.

Studies show that only the largest of eruptions, those that produce a stratospheric dust cloud, produce major climate change. The reason seems to be the Mediterranean's extreme sensitivity to climate change discussed in the previous chapter. The implication, however, is that with volcanoes, the greatest threat may not be the one on your doorstep.

SLEEPING GIANTS

For the modern inhabitants of Naples in Italy, the threat most certainly is on their doorstep. The chaotic, dazzling city throbs under the sleepy shadow of Mount Vesuvius. The first-century Greek geographer and historian Strabo had recognised the mountain as a volcano, but it took the terrible eruption of AD 79 to show the world what it could do. That eruption left the towns of Pompeii and Herculaneum entombed in volcanic ash and mud, sediment that within a few years would lure Roman farmers back to its deadly slopes. The temptation was the fertility of the volcanic soils, which offer high water retention, good drainage, easy cultivation and a rapid release of essential nutrients. However, the soils are also readily eroded, particularly on the steeper slopes, and can go from sterile ash to cultivable soil and back to sterile ash within centuries. Nevertheless, today, olive orchards and vineyards drape the lower moist slopes – the volcano even has its own Vesuvius wine – and Neapolitans continue to cultivate them in the full knowledge that they are flirting with disaster. It's a risk worth taking. The volcano last spewed out lava and ash in 1944, as Allied forces were pushing north through Italy on the way to far more lethal hills like those at Monte Cassino.

Although its classic cone shape makes Vesuvius the most famous volcano in Italy, it isn't the greatest threat to Naples. Tucked between Naples and the Tyrrhenian Sea is the Campi Flegrei, the fiery birthplace of myth and legend. It was here that Hercules (the Roman version), Ulysses and Aeneas faced some of their greatest challenges; it was here that the ancients placed one of the main entrances to Hades; and it was here that the Earth was shaken by one of the biggest eruptions known in the Mediterranean.

Campi Flegrei is Greek for 'Burning Fields' and it marks the site of a giant crater, twelve to fifteen kilometres across. It first blasted into existence 35,000 years ago, when one or more eruptions spewed out a vast blanket of pumice that today extends over 30,000 square kilometres. From out of its throat came at least 200 cubic kilometres of magma – the equivalent of 200,000 solid stone Empire State Buildings. Some geologists speculate that the global impact of this eruption might have been the straw that broke the camel's back for the Neanderthals, and might have led to their replacement by the incoming modern humans. The blast caused the area to collapse into the broad dish-shaped caldera that now cups the western part of the Bay of Naples. Twenty-three thousand years later, the Campi Flegrei blew again to throw out the famous Neapolitan Yellow Tuff that gave the Romans their pozzolanic cement (as discussed in Chapter 3). About six kilometres across, the crater from that eruption today lies underwater in the Bay of Pozzuoli.

A repeat performance of any of these blasts would be difficult to contain. There are about one million people at risk in Naples proper. Across the bay the beautiful volcanic island of Ischia is home to 50,000 people and the population swells in size with the hordes of summer tourists. How could such a throng be evacuated? Clearly the Campi Flegrei has the potential to produce a disaster on an epic scale. It is for this reason that the Burning Fields are now under constant scrutiny, monitored by a range of volcanic surveillance equipment – seismometers to pick up the vibrations of small quakes, gas sniffers, and sensors to detect tilting ground – all linked via satellites to feed data to the observatory for prognosis. The plan is that if the underground network of fissures below the crater fill with magma and push towards the surface, there should be forewarning. The only trouble is, the area is on the move all the time, sinking and swelling.

If the Campi Flegrei blows its top, the immediate devastation of the Naples bay may not be the only result. Close to Pozzuoli lies an enormous yellow-white crater called Solfatara. Here the air is thick with the tell-tale 'rotten eggs' smell of the sulphurous gases that billow out from cracks, a concoction that stems from the mixing of groundwaters and magma gases several kilometres beneath the surface. If the next eruption launches this sulphur stew high into the atmosphere, the resulting dry fog could put the whole Mediterranean into the fridge.

Far to the south there is another slumbering giant – Mount Etna. Forty

kilometres across and more than three kilometres high, it is Europe's largest and highest active volcano. Little wonder that this mountain has been home to gods and giants all through antiquity. The Greeks had a temple to their fire god Hadranus at the foot of Etna, and the deepest part of the Underworld was located here as was the metalworks of the Greek god Haephaestos, and his Latin equivalent, Vulcan, where his assistants, the Cyclops, crafted the bolts of thunder and lightning with which Zeus controlled the universe. The Cyclops themselves are traditionally believed to be preserved in stone as volcanic islets just offshore of the modern fishing village of Aci Trezza, north of Catania.

Mount Etna towers over eastern Sicily, dominating the lives of more than two million people, most of them relying on the volcano to maintain the fertile soils that support countless vineyards and olive groves. At first glance,

Vesuvius, the quintessential volcano, looms above Naples, its twin peaks both a lure and a menace. It provides some of the best orchards and vineyards in Italy and yet it has a population of three million within its eruptive range. Despite this potential for geological carnage, an even greater volcanic threat lurks beneath the shallow waters of the Bay of Naples.

'slumbering' seems the wrong description, since the summit vents of the volcano are almost continually exploding and degassing, and every few years lava flows are ejected from the lower flanks, engulfing farmland and threatening towns like Zafferana and Nicolosi with destruction. But in truth someone would need to be either extremely unfortunate or careless to be killed by Etna's effusions, either getting too close to the showers of rocky debris thrown from one of the four summit craters (today, tourists are kept a safe distance away), or being close to the lava front when it hits a patch of wet ground and explodes out molten shrapnel. Yet, despite the regular outpourings of lava and violent fire-fountains that every few years stir up a flurry of activity among the villagers, the emergency services and the world's media, throughout the last six and a half thousand years of human history Mount Etna has been relatively restrained.

Before then, around 6,700 years ago, Mount Etna blasted a massive hole in its side (exactly as Mount St Helens in Washington State, USA did when it famously exploded in 1980). The result was the creation of a deep, cliff-bounded amphitheatre called the Valle del Bove that scars the eastern flank of the great edifice. The hole is about five kilometres across and eight kilometres long, suggesting that 12 cubic kilometres of material was flung out. Such sudden enormous collapses of the sides of volcanoes, and the large landslides that result, are now recognised as part and parcel of the normal 'growing up' of many volcanoes – it's what might be called their adolescence. It mostly affects those lying near the coast or on islands, where the unsupported seaward sides have the opportunity to shift laterally during the build-up to eruption. As well as Mount Etna, and volcanoes further afield in the Hawaiian islands, the volcanic Canary islands – Tenerife, Grand Canaria and La Palma – show the deep scars of past violent slumps.

What actually triggers these giant flank collapses isn't clear but, yet again, a changeable climate seems to be at the heart of it. Rapidly rising or rapidly falling sea levels have the potential to destabilise the coastal flanks, and the timing of volcanic eruptions in the Mediterranean region indicates that volcanic eruptions were most prevalent when past sea levels were on the move, but it also seems that high water pressures are needed within the volcano to lubricate the moving flank. In this regard, planetary warming after the last Ice Age may have had a part to play, melting ice-capped summits and increasing rainfall to leave volcanic edifices saturated with water. So it is not unexpected then that the greatest eruption at Mount Etna happened around 7,000 years ago, a time when southern Europe was enjoying its highest temperatures, wettest weather and still rising seas.

SEISMIC FAULTS, SACRED SANCTUARIES

The hot springs and rich soils that volcanoes offer are, it seems, sufficient recompense to balance out the occasional catastrophic mayhem they cause when the largest of them blow their tops, but what about earthquakes, that other geological weapon of mass destruction that Earth wields frequently across the Mediterranean realm? What positive benefits did Poseidon's seismic rumblings have on the ancient world? After all, geologists now know that the Mediterranean region is criss-crossed by a giant cobweb of earthquake faults, lines of geological violence that mainly concentrate along the edges of the tectonic plates, but also splay off into their interiors. The result is that virtually anywhere in the Mediterranean is close to a fault that could rupture. Faults are important pathways for water to get to the surface and reliable water is an essential ingredient for any city, which is one of the reasons why most of the major cities of the ancient world lie on or close to faults.

There is, however, another intriguing possibility: that some ancient cultures deliberately chose to build their most sacred sites *directly on* earthquake faults. One particular culture that seems to have been prone to this was the ancient Greeks. Many of their sanctuaries – those vital conduits to the gods – are located on faults, in some cases the most sacred temples actually straddling the fault line. At ancient Knidus in south-western Turkey, the famous round temple of Aphrodite – popular in Classical times as displaying the western world's first monumental nude sculpture – is perfectly bisected and displaced by a fault. Nearby the same fault rises as a spectacular sheer wall, thirty metres high, at the base of which a temple to the goddess Demeter was apparently located, and into which have been chiselled rectangular niches to house votive offerings. Further north at Ephesus, the ancient Crevice Temple – an oracular temple of Apollo dated to around 400 BC – is so called because it is precisely positioned above a natural fissure in the limestone. A kilometre or so away, that fissure develops into a steep fault surface adorned by rock-cut shrines.

Quite to what or to whom ancient devotees were offering prayers at these sites remains uncertain, but in many cases it would seem natural to direct them to the gods of the underworld, and perhaps the fault lines were seen as passages leading down there. There is, however, another possiblity: that the devotees weren't there for what was going down, but what was coming up.

Ancient chronicles, like those of the Greek historian Plutarch, gave tantalising clues to mysterious processes at work in nature. He wrote about the most renowned oracle in the Classical world, in the Temple of Apollo at

The spectacular surface of an earthquake fault forms the backdrop to a temple sanctuary at the ancient Greek city of Cnidus in western Turkey. Rock-cut shrines adorning the near-vertical fault wall suggest that these natural fissures in the earth may have been the object of veneration in ancient times.

Delphi, on the slopes of Mount Parnassus. Here, in a basement room, a priestess would hang precariously above a chasm in the rock, until the god possessed her and she delivered his prophecies. Occasionally, she would be overcome and die. Plutarch reckoned that vapours came from the rocks below – and he was right.

For many centuries, into modern times, the story was confined to legend and hearsay, for no archaeological excavation could reveal the legendary chasm. Then, in 1999, geologists discovered that the Temple of Apollo lay directly on an earthquake fault and that hallucinogenic gases could be detected from within it. This limestone contains bitumen, a petrochemical that produces underground oil and natural gas. Friction along the fault line as the rocks rub together heats the bitumen, making it vaporise. The vapours bubble up through the groundwater and rise through the fault line and cracks. They end up in the underground chamber where the oracle gave her prophecies – validating Plutarch's idea of the vapour coming from the rocks. Recent analysis has found the gas contains small amounts of the drug ethylene, which is highly intoxicating. The so-called oracle, dispensing divine wisdom, was actually as high as a kite.

At Delphi, inhaling could seriously damage your health, but in a sacred city at the opposite end of the Greek world it spelled instant death. Hierapolis, in what is today western Turkey, was a Greek city that reached the height of its development as a Roman therapeutic centre in the second and third centuries AD. One of the most sacred buildings in ancient Hierapolis was a temple to Apollo. After nearly two thousand years and numerous earthquakes, very little of it remains, but the basement chamber that the temple foundations rest on survives along with its ominous reputation. The chamber was called the Plutonium. It was a sacred shrine to Pluto, the Roman god of the dead and the underworld, and the ancients believed it led straight down to Hell. As Strabo reported: 'Any living creature that enters will find death upon the instant.' Apparently, even bulls collapsed and died. The only humans that could endure the poisonous fumes were the Galli, eunuch temple priests who would venture underground to impress the sightseers, holding their breath for as long as possible to avoid unconsciousness and death. So thick were the vapours that Strabo claims it was impossible to see the floor. The dense, misty gas was carbon dioxide, a heavy vapour which when released from the bubbling spring water would pool in the basement of the temple, forming a suffocating mist.

Today the 'Sacred City' is a world heritage site that is home to one of the most remarkable geological landscapes in the Mediterranean. It's a dazzling fairyland of bleached limestone terraces which the Turks call 'Cotton Castle'

Opposite: The dazzling white cascade of travertine rock pools at Pammukkale in western Turkey are one of the Mediterranean's most stunning natural wonders. Hardly surprising that the Turks have dubbed this geological fairyland 'Cotton Castle' and designated it as a world heritage site.

(Pammukkale) – one of the world's natural wonders, brought about by a common earthquake fault. Groundwaters saturated with the dissolved ingredients of limestone are forced by heat in the hot rocks below back up to the surface, travelling up through the cracks in the rocks created by the fault lines. At the surface the carbon dioxide dissolved in the water bubbles out as gas, forcing the calcium carbonate to re-form. The water seeping from springs makes its way down towards a cliff edge and, as the water surges over the edge, a coating of calcium carbonate is deposited on the terraces. This slow, relentless deposition has been going on here for at least 14,000 years, plenty of time for the build-up of limestone to form these spectacular steps.

Since Roman times, the city's mineral-rich waters have attracted visitors in search of good health. Modern tourists swarm into Pammukkale in their coachloads, disembarking just long enough to bathe in the spectacular cascading water pools and frolic in the 'Sacred Pool' before heading back to their hotels four hours' drive away on the Aegean coast. The Sacred Pool is a natural spring that emerges among the ruins of the former city, and its waters are said to help with circulatory problems, high blood pressure, rheumatism and a range of other disorders. Yet the floor of the pool is littered with the ancient fallen ruins of a city that was devastated when the fault beneath its feet moved. The fault cuts its way through the ruins of the

The Sacred Pool in the heart of the geological fairyland of Pammukkale occurs where the warm waters that feed the terraced waterfalls emerge as springs. The waters surge up from an earthquake fault that cuts through the heart of the ancient Greco-Roman city of Hierapolis and the ruins of a past destructive quake litter the pool. Although the waters are renowned for their apparent therapeutic benefits, they bring gases like carbon dioxide which in antiquity were infamous for the lethal effects.

city – great walls that have toppled with shaking, streets that are bent and displaced by sudden jerks on the fault. The line of the fault leads to the Temple of Apollo and its lethal underground gas chamber, and all around is evidence of the damage that past earthquakes have done to the city – and yet, a couple of hundred metres from the Sacred Pool, is the only sign of its hidden secret – a fenced-off area where a spring continues to leak the noxious fumes pooling below in the Temple of Apollo's underground cavern. But the gas can be restless; some escaped in 1965 when a local earthquake caused the ground to open slightly.

The growing indications from Delphi and Hierapolis, as well as from Ephesus and Cnidus, are that the ancient Greeks did indeed deliberately place their main sanctuaries on earthquake faults. Whether this was because waters and gases emerging from them had therapeutic properties or provided essential elements of ancient rituals is not clear. Delphi and Hierapolis were both oracle sites and there were over three hundred other known oracle sites scattered across the ancient Mediterranean world, most concentrated in Greece and Turkey. Many involved priests drinking from springs or inhaling vapours from the rock. So were these too the result of gases rising up earthquake faults?

■ DESIGN EARTHQUAKES

At many of these sacred sanctuaries the seismic faults that were the subject of veneration and wonder would prove their undoing. Both Delphi and Hierapolis were destroyed and rebuilt several times after earthquake destruction, and seismic jolts plague them to this day. In fact, the Aegean region of Greece and western Turkey is the most earthquake-prone part of the Mediterranean, being struck regularly by damaging shocks. Earthquakes appear to have claimed as victim four of the wonders of the ancient world: the Colossus of Rhodes in the third century BC, the Great Temple at Olympia that housed the cult statue of Zeus in the sixth century AD, the Mausoleum of Halicarnassus in the thirteenth century AD, and the Lighthouse at Alexandria in the fourteenth century AD. Many of the ancient buildings and temples that are still standing show apparent signs of seismic wear and tear. At the fifth-century BC Hephasisteion (Theseion) temple below the Acropolis in Athens, for example, many of the individual 'drums' that make up the marble columns are misaligned. A Greek geologist fifty years ago proposed that this misalignment was probably from the rocking during seismic jolts, suggesting that Athens was not as immune from quakes as the last few centuries of quiescence had implied; a damaging quake that

Offset column drums in the colonnade of the 5th century BC Temple of Hephasisteion (Thesion) in Athens were attributed to rocking caused by earthquakes long before a damaging quake which struck the city in 1999 highlighted that although the Greek capital was not frequented by major shocks neither is it immune from seismic risk.

A displaced column drum in the inner colonnade of the Propylaea on the Acropolis in Athens was produced when an explosion in 1640 rocked the building, a reminder that much archaeological damage, past and present, is down to human actions.

struck Athens on 7 September 1999 was an even sharper jolt to that notion of immunity.

On the Acropolis itself there are signs of dislocation and breakage that some attribute to earthquakes in the twelfth or thirteenth century and one in 1705. Interestingly, the most obvious distortion, a displaced column drum in the inner colonnade of the Propylaea, is the result of a massive man-made explosion in 1640, and a later explosion in 1687, while much of the damage in the Parthenon was the result of an explosion during the Ottoman–Venetian war when the great temple was used to store gunpowder. Thus at least some of the distortions of the Acropolis buildings usually attributed to earthquakes are the result of human interventions. It is a cautionary reminder that earthquakes are just one of many forces that take their toll on ancient structures.

There is evidence that the ancient Greeks not only believed they understood the cause of earthquakes – according to Aristotle, underground gas trapped in subterranean caves – but they built their structures to take account of them. Despite the Classical ruins that litter the region, ancient Greek monuments were surprisingly hard to knock down. Columns, for instance, for all their apparent precariousness, are not easy to topple. Most ancient Greek columns are made of a series of cylindrical limestone or marble drums piled on top of each other; when an earthquake causes a column to rock the drums shift or jostle about, taking up some of the motion and making failure of the column as a whole less likely. Strengthening measures such as clamps and dowels were also put in place between column drums and between blocks in walls to hold the structure together. Often these 'ties' were made of hard iron or wood protected by a lead coating that was flexible and reduced corrosion; metal reinforcements on the Parthenon consisted of sulphur-free iron, also to avoid corrosion. At times it was this metal that would, literally, lead to their downfall, as in later centuries many temples were plundered for their iron and lead reinforcements, their columns and walls being toppled in the process to leave ruins that often would be mistaken by later archaeologists as signs of seismic destruction. Another way that ancient builders discovered to dampen the effects of ground shaking was to insert rows of bricks or timber into mortar walls, a practice first noted in the ruins of Akrotiri and still a popular feature of traditional houses in Greece. Finally, the tendency for walls to shift sideways during quakes, allowing roofs to cave in, was often remedied by the construction of stone buttresses to help support the wall.

Many of the specialist design features that the Greek architects inserted into their buildings became common practice for later cultures. In Byzantine

Previous page: The magnificent Haghia Sophia has dominated the skyline of ancient Constantinople and modern Istanbul for fifteen hundred years. With the city lying on one of the Mediterranean's most dangerous earthquake zones, it was designed and built to withstand the region's quakes but suffered three seismic collapses. With a future earthquake disaster looming for the city, one of the architectural wonders of the world is back in the geological firing line.

times, horizontal brick layers would become standard architectural design in towers and fortifications, such as the city walls of Constantinople (Istanbul), being both functional and decorative. As well as building to combat earthquakes, architects began to try to simulate their effects – a forerunner of modern earthquake engineering. One such architect was Anthemius of Tralles who, along with Isidore of Miletus, built one of the crowning monumental achievements of the ancient world: the Emperor Justinian's church of Haghia Sofia in Constantinople. To examine the effects of violent gas escape, Anthemius erected a system of cauldrons and pipes in his home in such a way that a jet of steam could be made to shake the floor of the apartment above. When he put his apparatus into operation, the upstairs occupants were so terrified of the shaking that they rushed into the street, convinced there had been a quake. Did this lead to success? No. Perhaps because gas escaping is not what actually causes earthquakes, the device was a poor test of the stability of the great temple and, barely twenty years after its completion, Haghia Sophia was severely damaged by an earthquake that struck in AD 557. Later seismic destructions brought parts of the great domed structure down again in 989 and 1346. Metal reinforcements and stone buttresses of the type introduced by the ancient Greeks were gradually added to keep the magnificent structure standing, though tourists can still see the signs of the stress and strain in the building's tilted floors, bulging walls and leaning pillars. Still, only three partial collapses in fifteen hundred years isn't bad.

IN THE FIRING LINE

Today, Haghia Sophia is in the firing line again. The great church, and the enormous modern city that surrounds it, lies along one of the Mediterranean's most dangerous earthquake faults. Turkey's North Anatolian fault is Europe's answer to the San Andreas Fault in California. This 900-kilometre crack in the ground cuts across northern Turkey from east to west, allowing the Anatolian crustal block to the south to move westwards with respect to Europe. Over the last sixty years, successive strands of this fault line have ruptured in large earthquakes, each event releasing stress on one part of the fault and passing it down the line to the next strand. The result is that, rather like a set of falling dominoes, the ruptures of the North Anatolian fault have moved steadily westwards – in 1939, 1942, 1943, 1944, 1957 and 1967 – from the comparatively sparsely populated parts of eastern Turkey to the industrial heartland of the north-west. Then, in August and November 1999, two of the strands just east of Istanbul that had yet to break ruptured in earthquakes that left over 35,000 dead, destroyed fifteen thousand buildings and cost $10–25 billion in damage. Earthquake geologists are convinced that the quakes have now added stress to the last remaining significant unruptured strand, the section of fault that lies in the Marmara Sea, directly offshore from Istanbul. If the pattern of the past continues, then Istanbul, a swollen urban sprawl of some ten million people, is due for a direct seismic hit in the coming years or decades.

It is difficult to overstate the threat posed to Istanbul. However, it is easy to overstate the effects that earthquakes have had in human history, where they have frequently been used by historians and archaeologists as convenient explanations for cataclysmic destructions and abandonments. In fact, earthquakes rarely wipe out entire cities, let alone entire regions. More often, seismic shocks leave cities as jumbles of some ruined, some damaged and some intact buildings, encouraging their inhabitants not to flee but to stay and rebuild their houses and livelihoods. It is a pattern that we see in modern earthquake disasters, and there is little sign that human nature was any different in the past. In general, the intensity of seismic effects weakens away from their source, ensuring that earthquakes are unlikely to wreak regional devastation. Typically, seismic disasters strike down one or two neighbouring settlements, but those a few tens of kilometres away are virtually unscathed. The size, or magnitude, of the earthquake determines the extent of the 'fallout zone'. The 1999 Turkish earthquakes struck only eighty kilometres from the centre of Istanbul, yet the death toll in the metropolis was comparatively light, approximately a thousand people. Even in great earthquakes, those with magnitudes exceeding 8 on the Richter scale, the destructive effects are limited to a hundred kilometres or so.

Nevertheless, some historians and geologists propose that around AD 365 a 'universal' earthquake shook the eastern Mediterranean. Accounts of seismic havoc at the time is described from Sicily to Cyprus and attributed to an enormous rupture of the subduction zone that allows the edge of the Africa plate to push deep below Crete and the Aegean. There is geological evidence for a momentous movement of the crust here in the form of the uplifted shoreline of western Crete, loosely dated to around the time of that famous quake. Other scholars are sceptical. Some historians argue that the Mediterranean-wide effects actually combine the damage of different earthquakes which struck in the years before and after 365. Also, archaeologists studying abandoned Roman harbours on Crete now believe that this upheaval actually took place over a century later, around 500. Nevertheless, the fact that the decade of the 360s sees a storm of earthquakes across the length and breadth of the eastern Mediterranean is intriguing.

If 'earthquake storms' can occur over decades, might that also explain apparent sudden bursts of devastation that affected the ancient world in a similar short period of time? Most controversial is the theory that an earthquake storm may have been responsible for the abrupt physical and political collapse of the Aegean Bronze Age world around 1200 BC. Some geologists and archaeologists point out that most of the ancient cities that fell at that time lie along the plate-boundary of the eastern Mediterranean and show signs of destruction typical of earthquakes. It supports a view that a storm of earthquakes successively 'unzipped' the plate boundary, so weakening the cities along the way that they were left militarily vulnerable, inviting attacks from opportunistic neighbours. Archaeologists remain to be convinced.

SPACE INVADERS

If home-grown destructive events are not enough, Earth is never short of space debris dropping in, with anything up to 40 million kilograms of space rock falling on our planet each year. Most of this rock we can see as 'shoot-

A circular lake in the Sirente national park in the Italian Apennines has been interpreted by some geologists as the impact crater from a meteorite strike dated around the fourth to sixth centuries. This Late Roman period is known to have been a time of enhanced meteor activity, leading some to suggest that cosmic fireworks may have contributed to the demise of the Imperial Rome.

ing stars' – small lumps of rock more properly called meteors. These meteors completely burn up in the sky due to the intense friction they create crashing through the gases in our atmosphere, but if a meteor is big enough and has the potential to survive the intense burning of the atmosphere, it can make it all the way down to the Earth's surface, where it lands as a meteorite.

Meteor streaks in the night sky have been a source of wonder since antiquity. Chinese imperial astrologers have an especially detailed record of these celestial fireworks from 100 BC to AD 1900, and this shows that there are occasional large surges in fireball activity lasting from decades to centuries. There are three surges that stand out over the last two thousand years: one from AD 300 to 600, one from 1000 to 1200, and one from 1300 to 1500. The surge of AD 300 to 600 spans the centuries over which the Roman empire declined, leading a group of cometary scientists to propose that the end of one of the world's greatest empires might have been caused by our planet being grazed by a shower of cometary debris.

Comets are icy bodies that normally reside in a 'cloud' of cometary

material in a distant corner of the solar system, but which occasionally venture out, pushed by collisions into long orbits that may take them close to Earth. Most comets that are seen travelling past Earth are between one and ten kilometres in diameter, but some are much larger. The break-up of these comets produces asteroids – irregular rocky lumps that generally lie in a vast belt of debris between the orbits of Mars and Jupiter. This 'asteroid belt' is the drifting fragments of larger bodies that have disintegrated and then been 'captured' by the pull of planets, particularly Jupiter, but it is constantly being nourished by the break-up of new comets.

Understandably, most attention has been focused on the potential for giant comets to collide with our planet. Kilometre-sized objects are expected to intercept our planet every 100,000 years, while 10-kilometre-sized monsters are likely to crash down every ten million to a hundred million years. Cometary encounters are now implicated in the five major mass extinction episodes that have affected life on the planet, and in a host of smaller mass-mortality events, though many geologists, perhaps most, still prefer homegrown terrestrial explanations. But far more frequent than these mega-collisions are the modest-sized rocky lumps that cometary scientists expect Earth to encounter every few hundred or few thousand years. Most of these are not expected to hit Earth directly, only to graze our atmosphere. Over the last 5,000 years, roughly the span of Mediterranean civilisation, theoretical estimates suggest that a few dozen 'bumps' with an energy of about 10-megatons of TNT are likely to have occurred. Such impacts are equivalent to the effect of the 30–50-metre large asteroid lump that exploded over Tunguska in Siberia in 1908, flattening 2,000 square kilometres of forest. More significantly, a handful of 100- to 1,000-megaton scale 'bangs' can be expected to have happened.

Our planet's orbit regularly takes us through the debris trails of comets. Earth's encounters with these give us the meteor shower displays which we can often see in dark, cloud-free skies, including the Perseid shower in mid August and the Leonid shower in mid November. Some of the most dramatic meteor displays occur in June and November when Earth passes the Taurid complex, a ring-shaped belt of fresh space debris believed to have formed only in the last 20,000 years by the break-up of a giant comet in our solar system. Most years, only light cometary dust from the Taurids strikes the upper atmosphere, but every few centuries Earth passes closer to the coarser centre of the Taurid asteroid trail, giving rise to spells of spectacular 'meteor storms'. And every thousand years or so the Earth's orbit takes it through the denser core of this ring of dust, causing larger rocky and icy lumps to enter our planet's atmosphere. When this happens, the twice-yearly bombardment becomes far more severe, raining down asteroidal debris over a period of a few decades on a scale comparable to a nuclear war. Some cometary scientists argue it was one of these more violent Taurid asteroid streams that Earth would have encountered during the period AD 300–600. The effects of a serious cometary dust-up would be dramatic; the millions of tons of meteorite dust would create a cooling dust veil, with the associated litany of crop failures, famine and disease that volcanic veils induce. The result is a so-called 'cosmic winter'.

If a shower of large asteroid fragments did rain down on Earth during late Roman times, then it probably did so during the remarkable decades of the mid sixth century AD. Tree-ring records from oak timbers from across northern Europe show virtually no growth for trees living between AD 535 and 550, suggesting a very severe climate, and the year 536 is the second coldest summer in the 1,500-year-long tree-ring record. Numerous Byzantine writers describe the extraordinary effects of a dust veil that enveloped the eastern Mediterranean between AD 536 and 540. According to one of them, Zacharias of Mytilene: 'The earth [at Constantinople] with all that is upon it quaked; and the sun began to be darkened by day and the moon by night, while the ocean was tumultuous with spray from the 24th of March till the 24th of June the following year . . . And as the winter [in Mesopotamia] was a severe one, so that from the large and unwonted quantity of snow the birds perished . . . there was distress . . . among men . . . from the evil things.' Byzantine records from a weather station at Baghdad confirm the extreme coldness and heavy snowfalls that occurred in the winter of AD 536/537.

The remarkable events of this period appear to have been widespread. Chinese chronicles similarly record great falls of yellow dust towards the end of AD 536, and severe hardships of famine and disease followed for the next few years throughout China, Korea and Japan. Evidence for cooling also stretches the length of the western Americas, from California to Chile, supporting the notion that this was a global event. For many geologists the signs of climatic cooling across both hemispheres indicate a large volcanic eruption, but while volcanoes in Iceland, Indonesia and Papua New Guinea have been implicated, no one has yet been able to find a convincing smoking gun.

Cometary bombardment is also being invoked to explain some of the other major environmental crises of the last few thousand years, including the Bronze Age collapse of the Mesopotamian and Egyptian Old Kingdom empires in 2300 BC and that of the Aegean Bronze Age societies of 1200 BC. Cometary scientists are also starting to implicate the impacts of celestial showers in more recent times, such as the cool, dry period of the early ninth century AD when there was major turmoil in northern Europe and the main Mayan civilisation collapsed in Central America. Some even see cosmic destruction as the driver of other events. Minor surges in Chinese fireball chronicles accord with some of the turning points in Mediterranean and world history, including the initiation of the Crusades against Islam, the English Civil War, and the American War of Independence allied with the French Revolution. Most archaeologists and historians, it should be said, are yet to be convinced.

A RISKY BUSINESS

Earth is a dangerous place to live, and it will continue to be so. The earthquakes, volcanic eruptions and cometary impacts that can destroy are simply the way that our dynamic planet, and its busy cosmic neighbourhood, works. Trapped between the might of African and European plates, for millions of years to come the turbulent Mediterranean region will continue to be haunted by a geology that has a devastating capacity to deliver both good and ill.

Geological violence, however, may be a driver for change. Human ingenuity and technological development, it seems, flourished in a challenging and tumultuous natural world. The Mediterranean region may be one of the most geologically restless places on Earth, but the fact that it has given rise to the first civilisations and to some of the world's most powerful empires suggests that humanity has not been cowed by such planetary violence. Geological turmoil was simply another challenge in a restless world of war, drought and disease. Natural catastrophes may have brought disaster to those in the firing line, but for others nearby there were opportunities. As with the battles against a changing climate, the constant need to adapt and overcome natural disasters has no doubt encouraged the exchange of peoples, ideas and technologies.

Such words are scant comfort, though, to those living in geological risk hotspots like Naples or Istanbul. In the Mediterranean, as across the globe, more and more people are crowding into cities, making them larger targets for seismic, volcanic or even cometary destruction. Incidences of destructive geological events today are probably no more or less than in the ancient world, and in broad terms they are happening where they have always done, but the growing urban centres, many built with poor quality housing, are ensuring the region is becoming more vulnerable to natural disasters.

It all leads to an increasing likelihood that in the twenty-first century humans will experience a scale of catastrophe that far exceeds anything the historical world had to endure, and yet, perhaps the greatest threats that we face are not from geology at all, but instead from a force that is shaping the Mediterranean faster and more dramatically than nature ever has. That force, of course, is humans themselves.

CHAPTER 8

UNCHARTED WATERS

The influence of the human race on the planet is now as profound as that of natural forces. During the last three centuries, the population has increased tenfold to 6,000 million; the world's cattle population (which does have a bearing on this) has risen to 1,400 million – about one cow per average size family. Some 30–50 per cent of the land surface has been transformed by human action; dam building and river diversion are commonplace. Urbanisation has increased tenfold in the past century. In a few generations people are exhausting the fossil fuels that were generated over several hundred million years. Energy use increased six-fold during the twentieth century and at present the release of sulphur dioxide by coal and oil burning is at least twice as large as the sum of all natural emissions. Fossil fuel burning and agriculture have substantially increased climatically important 'greenhouse' gases in the atmosphere. Modern human actions release copious cocktails of substances into the environment, many of which are toxic to life directly, while others, such as the chlorofluorocarbon gases that eat away at the 'ozone hole', are toxic to our protective environment. Tropical rainforests disappear at a furious pace, releasing carbon dioxide and strongly increasing species extinction by a thousand- to ten thousand-fold. More than half of all accessible freshwater is now used by people, but much of the rest is being contaminated by mining and farming wastes. Coastal wetlands are disappearing, with the loss of 50 per cent of the world's mangroves. If this wasn't enough, in the last century mechanised fishing increased the annual harvest of fish from the world's oceans from 5 to 90 million tons per year, removing more than 25 per cent of the primary production of the oceans in the upwelling regions and 35 per cent in the temperate continental shelf regions. All this environmental carnage has been caused by only 25 per cent of the world's population – the industrial, developed, mainly Western, world.

Some scientists who study Earth systems – how our planet functions and maintains itself in a fit state for life – reckon that as the effects of human activities on the global environment have escalated to an unprecedented degree over recent centuries, we have entered a new geological age. Having lived for the last ten to twelve thousand years in a warm but changeable age called the Holocene – a response to the climatic after-effects of the Ice Ages – we are now in the Anthropocene Age: the age of humankind.

One contender for the start of this new epoch could be the late eighteenth century, when the global effects of human activities became clearly noticeable. Chemical analyses of trapped air inside polar ice show it is around this time that growing global concentrations of carbon dioxide and methane – the gases that seem to underlie the current global warming – are found. A precise year for the start of this 'human age' could be set as 1784, the date of James Watts's invention of the steam engine which effectively marked the start of the great Industrial Revolution.

However, many argue that if we are in a new human age then it began far earlier than the late 1700s – in fact it was with the transition of people from hunter-gatherers to farmers thousands of years ago, when primitive agriculture began to change the Earth's climate. Some geologists point the finger at the ancient Mediterranean and Near Eastern civilisations, which apparently deforested their region and turned it dry. Such a criticism seems unfounded given our realisation that these early farming societies were struggling to overcome a drying world, responding to forces beyond their control.

Nevertheless, the actions of these first farmers probably did have serious and permanent effects on the world so subtle that we are only today picking them up. In particular, the rise of rice cultivation in the ancient world may have contributed to the present glut of methane that is helping warm our atmosphere. Rice cultivation flourished in the warm, wet climate of the early Holocene, pumping out methane. Later, with the decline in solar heating and consequently in monsoon rains, farmers were forced into artificial flooding to create wetlands. While natural wetlands went into decline around the world, the artificial wetlands boomed, ensuring that the methane-producing plants were still able to grow. Animal domestication too may have contributed to our present methane problem. Hoofed animals – cattle buffalo, sheep and goats – cannot digest much of their food until it has been fermented, and the bacteria that carry out that fermentation produce methane as a by-product. Domestic animals contribute 15–20 per cent of the methane output from our agricultural and industrial activities, with cows being the worst single offender, accounting for around 70 per cent of the methane emissions from domestic animals.

Large increases in carbon dioxide in our atmosphere may also be due to agriculture, or at least to the deforestation that accompanied it. Before the Industrial Revolution, all civilisations relied on wood for fuel, and it is estimated that the loss of forest – slowly from around 8,000 years ago and more rapidly from 2,000 years ago – released enormous quantities of carbon into the atmosphere. Intriguingly, there are periods of a few hundred years during the last 2,000 years when the carbon dioxide levels have suddenly dropped. Since humans were the main agent causing the general rise through deforestation, the occasional reverses in the upward trend may relate to sharp reductions in the human population. Wars and drought-induced famines are probably too short and too limited in extent to be the culprit, but plagues are powerful enough to have an impact on the global population. Some climate scientists are now making the connection between plagues that ravaged the world in the sixth and fourteenth centuries AD and the anomalous carbon dioxide dips. As people died, farmland was abandoned and forests recovered, drawing back carbon dioxide from the atmosphere.

The doomsday scenario raised by some in the scientific community is that without major catastrophes like an enormous volcanic eruption, an unexpected epidemic, a large-scale nuclear war, an asteroid impact or a new ice age, humankind will remain a major geological force for the foreseeable future. Because of the anthropogenic emissions of carbon dioxide, climate may depart significantly from natural behaviour over the next 50,000 years. With this rise in humankind, our impact on the planet has risen to the global scale and we have entered uncharted waters in terms of how climate and the environments that sustain life on Earth will respond to the changes being imposed on them. As a leading American environmentalist put it: 'We are as gods and might as well get good at it.'

■ A RUINED LANDSCAPE?

This concern that we are living beyond our planetary means is not new. Tertulian, a third-century AD theologian, lamented that:

> Surely it is obvious enough if one looks at the whole world, that it is becoming daily better cultivated and more fully peopled; all places are now accessible, all are well known, most pleasant farms have obliterated all traces of what were once dreary and dangerous wastes; cultivated fields and subdued forests, flocks and herds have expelled wild beasts; sandy deserts are sown, rocks are planted, marshes are drained; and where once there were hardly solitary cottages, there are now large cities. No longer are islands dreaded, not their rocky shores feared; everywhere are houses and inhabitants. Our teeming population is the strongest evidence; our numbers are burdensome to the world that can hardly supply us from natural elements; our wants grow more and more keen, and our own complaints more bitter in all mouths, whilst Nature fails us in affording us her usual sustenance. In very deed, pestilence and famine and wars and earthquakes have been regarded as a remedy for nations, as a means of pruning the luxuriance of the human race.

Tertulian was writing just as the environmental and geological pressures were starting to bite on the great Roman empire, but the idea that humans were already ruining the Mediterranean landscape in ancient times has been a persistent and popular notion. The Victorian travellers and writers who embarked on the Grand Tour that took them on a nostalgic trip through the Classical world were disillusioned to find the 'barren' lands of ancient Rome and the 'bare' hills of Pericles' Attica. Brought up on seventeenth-century Renaissance poets and Baroque painters who set their depictions of antiquity against the lush backdrop of northern Europe, they assumed that the landscape of the Mediterranean must have 'gone bad'. A myth developed of a Mediterranean Eden, clothed in magnificent forest, cut down thoughtlessly by an irresponsible populace to feed their immediate needs of houses, ships and fuel. With the venerable trees replaced by the voracious goat, the sorry tale ended with the rich soil being lost to the rivers and sea and the bare carcass of the land exposed. Successive generations of peoples since the Classical heyday – the Arabs, the Venetians, the Ottoman Turks – were seen as unfit guardians of what was once a veritable Garden of Eden.

No doubt it is a view that many visitors from northern Europe have today as they arrive in the dry heat of the summer to be confronted with the fawn hues of a rocky landscape of sparse trees and shrubs (often burnt trees and shrubs). Compared to the green rolling hills and woodlands many left behind, it must seem a fragile natural world clinging on to existence, but the reality is very different. Mediterranean countries have a richer diversity of plant life than the rest of Europe. Many common species do not occur north of the Alps; some are of African or Asian affinities. The diversity arises because plants have had to evolve to make the best of the environments into which the accidents of climate and geology have thrust them. Probably the most versatile plants are the thick-leafed evergreen shrubs and small trees called 'maquis': long-lived, deep-rooted, relatively palatable and combustible, but not killed by wood-cutting, burning or browsing, they are ideally adapted to the capricious Mediterranean environment. The 'noble forests' of Classical antiquity were probably of this same 'savanna' woodland, pinched by drought, browsing or cold into stunted oak and pine trees scattered as stands amid grasslands or scrub. Burning in particular is essential for success here. Some trees and plants are insulated against natural fires with fire-proof barks, while others sprout or germinate after the fire has swept through, producing flammable resins and oils to encourage the flames that will regenerate them and set back their less fire-adapted competitors.

However, fire is thought of as a human-inflicted misfortune of Mediterranean lands. It is true that arson is common practice in land disputes, with developers and speculators buying burnt agricultural or rough land at bargain prices, but it is also true that over thousands of years the Mediterranean environment had adapted itself to rely on natural conflagrations. More significantly, the suppression of fire only leads to the build-up of combustible debris to levels that ensure when a wildfire does take hold, it quickly turns into a vast uncontrollable inferno. Yet when fires do sweep across the tinderbox of parched vegetation, media headlines and authorities bemoan the widespread 'destruction' of pristine wilderness. Often the concern is for the threat to affluent houses or villages that chose to site themselves on the edge – or worse, in the middle – of this fiery landscape. Nevertheless, the present policy in most Mediterranean lands is to 'control' fire, through anti-fire legislation and campaigns to educate people against their accidental ignition. Fires of course have not gone away, and if we are serious about wanting to maintain the natural environment in Mediterranean lands they should be promoted. Preventing people building houses in fire-prone areas, and deliberate controlled burning to remove the build-up of flammable litter, are two important ways that this could be done.

Fauna too have adapted to this changeable world. The most prevalent animal is the goat, which is much maligned for its efforts in laying bare the barren rocky skeleton. As they scour the craggy uplands for meagre sustenance, it seems that they strip away the soil and eat young plant growth, but the soils on the abundant limestone plateaux and hills are naturally thin and patchy, yet still offer far more sustenance to animals than do the small, better vegetated pockets of sediment trapped in the upland valleys. Soluble phosphate, the principal vegetational contribution to the calcium compounds from which growing animals make bones, is between two and four times higher on limestones than on most other rock types, and only the limestones offer adequate trace elements for animal growth, such as copper and cobalt. Goats, long seen by European incomers as inferior to the sheep and cattle of the lush northern pastures, are, along with deer, the perfect animals to root out the varied edible plants that prosper in this natural rockery. They nibble too at the woody shrubland, consuming much of the plant debris that would otherwise fuel fires.

Much of the Mediterranean region has the appearance of a land laid bare. From Plato, to Pliny to Tertulian, Classical writers have complained of a landscape that is haemorrhaging the soil that is its lifeblood. Erosion, of course, is a natural process that has been going on for millions of years, accelerated in places where the geological upheavals of the past have pushed up mountains and thereby captured the moist Mediterranean winds. Today, erosion rates, as measured by the amount of sediment emerging from the great rivers and deltas, are still greatest in those areas that are being uplifted by tectonic movements and shaken by earthquakes, but there is little doubt that human actions, with fire, land clearance and agriculture, have the potential to move material more effectively than natural actions. Since the Neolithic farmers arrived from the east, the slopes of the Mediterranean appear to have been shedding their debris in greater amounts. It remains questionable whether the increase is from human mismanagement of this natural resource or from a losing battle with an unpredictable climate which at times encourages land degradation. What is clear is that much of the sediment liberated during the earliest times of human clearance – the Bronze Age – was washed down into the valleys and coastal plains to provide the growing fields for later societies and empires. Great cities like Troy and Ephesus were located to exploit the growing fertile plains which centuries later would engulf them. Furthermore, some of the barest landscapes of the region – the deserts, karst lands and ravines and gullies of eroded 'badlands' – are likely to have been similar in the past, scarcely mentioned by travellers who were more interested in human landscapes.

The natural history of the Mediterranean is a story of resilience, not decay. It is the history of a land and people adapting to a world that switches seasonally and annually and over centuries and millennia, between cold and hot, wet and dry. It is a land very different from its neighbours, one in which there is nothing abnormal about violent natural events – deluges, fires, earthquakes, volcanic eruptions – which can change the landscape overnight. The ups and downs of climate and geology have acted together with the ebb and flow of civilisations and peoples to shape the present-day Mediterranean landscape – but now, it would seem, even the natural robustness of the Mediterranean world will be tested as humans become the main geological force on the planet.

■ INTO THE LOOKING GLASS

Within the last century, the Mediterranean has been transformed. A region that was predominantly poor, over-populated and agricultural is developing into one of the most prosperous regions on Earth. Tourism, a feature of the region for over a thousand years, has made it the most popular holiday destination in the world. In the Mediterranean, tourists swell the populations of cities and coastlands, roughly doubling the 130 million or so people that live around its shores. Here, agricultural activity has been concentrating on the plains where machinery and irrigation can readily be employed, creating pressures on resources almost as acute as those in underdeveloped countries: the demand for water by agriculture and cities is coming to exceed supply.

In the mountains and countryside there has been the reverse. A century ago around 30 per cent of the region's working population were involved in agriculture; today it is around 3 per cent. Cultivation has been abandoned and fewer flocks and herds move from the plains in the winter to the mountains in the summer. As the twentieth century progressed, the diversity of landscapes diminished, many being replaced by the dull uniformity of plantations, concrete towns and golf courses. In the African Mediterranean the move towards industrialisation and tourism has been less dramatic, and here agriculture is expanding to support a growing indigenous population, but all under the shadow of drought.

In a world that is warming up, the sensitive Mediterranean will be one region where the effects of climate change are likely to be felt most acutely. Globally, by 2025, it is estimated that out of a total population of 8 billion people, around 5 billion will be living in countries experiencing water stress, and many of these will be around the Mediterranean. In the case of Spain, 54 per cent of its population now lives within fifty kilometres of the coast. In the Spanish coastal plains industry, tourism and expanding cities compete with agriculture for water. To put this into perspective, in Spain people use about 300 litres per day and a five-star hotel uses about 500 litres per occupant per day. Golf courses, by comparison, have come to 'need' 100,000 litres per day per hole. The water drunk by agriculture dwarfs nearly all other uses. A cultivated plain a hundred square kilometres in size can be expected to be irrigated with 20,000 million litres of water per year. The same quantity of water should provide for the drinking and washing of a city of 180,000 people, or for a hundred five-star hotels with 1,000 guests, or thirty eighteen-hole golf courses.

Around the Mediterranean, rivers are drying up, drained by irrigation or dammed. Many of those continuing to flow are polluted by farming residues. Some of the demands on the surface water is coming from tourism, but most is from the new agriculture. The face of agriculture across the region has changed dramatically. Traditional mixing of crops has given way to oranges and other fruits, which are more lucrative but more unstable and water-demanding. Irrigated fruit and vegetable plantations have turned many a fertile plain into a sea of plastic hothouses. In Spain, water-demanding crops like maize and alfalfa (lucerne) have decreased, but these have been more than made up for by increasing irrigation of vines. La Mancha in Spain holds the record for the greatest concentration of vines in the world, vine-growers here irrigating 1,300 square kilometres of vineyards. It is impossible to meet the extra demand with surface water, which in Spain provides only 25 per cent of the needs of farming, 2,000 million cubic metres per year.

Previous page: Increased flooding of Venice over recent decades was mainly the result of geological subsidence, but future global warming of the planet is threatening to engulf this remarkable Italian city with rising seas and more frequent storm surges.

Instead, around 37 per cent of Spain's agricultural thirst is slaked from the ground, an amazing 5,000 million cubic metres per year.

Until recently, much of the region's groundwater has been drawn from shallow depths. Extraction rates have rocketed in recent decades, often becoming more than the rate at which rainfall refills the underground storage systems. In many areas, the water table has dropped by a few hundred metres. To make up for the sinking water, seawater gets drawn in to many coastal plains. About one-third of the irrigated land in Spain is within reach of seawater, meaning that the risk of salty water poisoning many crops is a growing problem. Citrus groves are the most sensitive crops to salinisation, and the first to be abandoned. The salts also corrode wells and pumps, and clog pipes with limescale. All of these effects are harmful, costly and difficult to remedy and all are likely to increase in the future. And the more water the ground provides, the more demand will rise to meet, and exceed, it. Surface water, for all its unpredictability, is easy to manage – reservoirs, canals or ditches are built to store and move it around. Groundwater, by comparison, is an invisible resource, hidden below ground, so it is never really known whether it is drying up.

Groundwater extraction has another unwelcome side-effect – the land of the coastal plain drops as water is removed from below. In a warming world of rising seas, coastal subsidence induced by groundwater removal presents a growing threat. Global sea levels are expected to rise by 20 to 90 centimetres in the next century, posing problems for many of the Mediterranean's largest cities that sit on the edges of low-lying muddy deltas. These deltas are already sinking: the Nile by about 2 millimetres per year and the smaller deltas of the Rhône, Po and Ebro even faster, at 3–10 millimetres per year. The sinking is from the slow compaction of soft muddy material underfoot, and in some cases from ongoing tectonic movements, but mainly it is from the groundwater being sucked out of these industrial and agricultural heartlands.

Venice is a prime example of a city that has been pulled down by its over-zealous removal of groundwater, which by 1970 had caused the city to subside by twenty centimetres. Although much of that has now stopped, there has only been a recovery of a few centimetres. The withdrawal of gas from giant gas fields just off the northern Adriatic coastline is threatening the hundred kilometres of coast from Venice to Ravenna – most of which is within a metre or two of sea level – with submergence. It all makes for a stormy future for Venice. At the beginning of the twentieth century, St Mark's Square was under water a handful of times a year, a novelty for its winter visitors; by 1990 this was happening forty times a year. The normal tidal range in the northern Adriatic is a metre, but only a 30-centimetre rise in sea level is needed to make St Mark's Square into a permanent paddling pool, so engineering work is currently under way to raise up the whole area. Over the longer term, an ambitious but controversial system of flood gates is under construction to protect the whole lagoon from the stormier seas that global warming forecasts for the future.

Around the Mediterranean, increased coastal flooding is likely to take its toll in a narrow zone where much of the region's infrastructure is concentrated. There is already a large population located in this coastal zone, and a number of large and rapidly growing cities like Barcelona, Athens, Istanbul and Tripoli sit at the coast. With the mid-twentieth-century pursuit of sun, sea, surf and sand apparently undiminished, the twenty-first century will see even greater tourism here, anything between 170 and 340 million visitors are predicted by 2025. This means that a huge tourist infrastructure already exists, or will be built, immediately adjacent to the coast.

Those who visit the Mediterranean region generally do so out of a genuine affection for its landscapes, cultures and people. Yet the visits that many of us make to its shores are too brief to appreciate the complexities of the environment that it struggles with. Its vibrant history, one of the magnets that draw us there, tells the story of the region's struggles with a capricious climate and a restless geology. It gives us clues about what is in store.